Living with the Chip

Living with the Chip

David Manners
Reed Business Publishing
Sutton
UK

and

Tsugio Makimoto
Semiconductor and Integrated Circuits Division
Hitachi Ltd
Tokyo
Japan

CHAPMAN & HALL

London · Glasgow · Weinheim · New York · Tokyo · Melbourne · Madras

Published by Chapman & Hall, 2–6 Boundary Row, London SE1 8HN, UK

Chapman & Hall, 2–6 Boundary Row, London SE1 8HN, UK

Blackie Academic & Professional, Wester Cleddens Road, Bishopbriggs, Glasgow G64 2NZ, UK

Chapman & Hall GmbH, Pappelallee 3, 69469 Weinheim, Germany

Chapman & Hall USA, One Penn Plaza, 41st Floor, New York NY 10119, USA

Chapman & Hall Japan, ITP-Japan, Kyowa Building, 3F, 2-2-1 Hirakawacho, Chiyoda-ku, Tokyo 102, Japan

Chapman & Hall Australia, Thomas Nelson Australia, 102 Dodds Street, South Melbourne, Victoria 3205, Australia

Chapman & Hall India, R. Seshadri, 32 Second Main Road, CIT East, Madras 600 035, India

First edition 1995

© 1995 David Manners and Tsugio Makimoto

Typeset in 10.5/12 Times by Saxon Graphics Ltd, Derby
Printed in England by Clays Ltd, St Ives plc

ISBN 0 412 61690 4

A catalogue record for this book is available from the British Library

∞ Printed on permanent acid-free text paper, manufactured in accordance with ANSI/NISO Z39.48-1992 and ANSI/NISO Z39.48-1984 (Permanence of Paper).

*For Sash, Gem, Toshiki, Yoshiko,
Nobuo and Akio*

Contents

Chapter 1 What does the chip do for me?   1

How 'killer' products emerge through microelectronics advances
... the surprise of transistor radios, calculators, PCs, faxes, pocket
phones, camcorders etc. ... artificial senses: sight, hearing, touch,
smell, taste and brain power

Chapter 2 The pioneers   16

How it all happened ... Edison's tube ... Shockley's transistor ...
Kilby's chip ... erratic early days in transistors ... entrepreneurs vs.
established corporations ... booms and busts ... effect on computers,
calculators, watches ... innovation followed by commoditization ...
collapsing prices ... early days in chips ... emergence of PCs

Chapter 3 How the chip became an industry   35

America invents an industry ... Silicon Valley gets started ...
Fairchild ... Intel ... memory chips and microprocessors ... missiles

and Moon rockets ... Japan mobilizes and catches up ... Europe intellectualizes and wilts ... Taiwan and South Korea emerge

What can a chip do? ... What will it soon be doing? ... impact on computers ... impact on communications ... free storage, transmission and processing ... merging of equipment types and information types ... superhighways and their effects

Moore's Law ... the Pentagonal Pyramid ... Makimoto's Wave ... the incredible shrinking transistor ... the Silicon Cycle ... growth trends to 2000 AD

The main chip players ... their origins ... their contributions to microelectronics ... their successes and failures

Great disasters – the Schlumberger takeover of Fairchild and United Technologies' acquisition of Mostek ... Great triumphs – Chips and Technologies, Intel, Samsung ... the advantages of the paranoid

A guide for the outsider in coping with the chip-makers ... how to invest and when to disinvest ... who to work for, who to avoid ... how to choose a chip supplier and what to expect ... how to be a canny consumer of chip-based products

Preface

Everyone nowadays knows that electronics goods get cheaper and better every year, but few people know why. Those who do know why can foresee the kind of future goods that will be made, when they will be sold and at what price. Those who don't know will fail to understand even today's products.

Those involved with electronics goods, from pocket telephones to pocket computers, from camcorders to fax machines, whether as entrepreneurs, investors, inventors, retailers or consumers are unable to make intelligent decisions without understanding why new or better products have appeared and what can be expected in the future.

The technologies that make these products and their continual improvement possible have cost countries and companies billions to develop, and their success, or lack of it, in electronics has caused the rise, or relative decline, of whole economic regions.

Yet all these capabilities are based on a single physical phenomenon – the incredible shrinking transistor, or the science of microelectronics. This science is in the hands of a relatively tiny bunch of people. The purpose of this book is to be a bridge between those people – the microelectronics community – and the general public.

Along the way we may upset the former by over-generalization and the latter by including too much detail. If this is so, we offer our apologies.

Acknowledgments

The authors would like to acknowledge the help of colleagues in preparing this book and to thank them for their time, ideas, advice and contributions.

From Reed Business Publishing:

Richard Wilson
Leon Clifford
Karl Schneider
Simon Parry

From Hitachi:

Hajime Yasuda
Dr Takaaki Hagiwara
Mitsugu Yoneyama
Jim Duckworth
Matthew Trowbridge

1
What does the chip do for me?

Some innovations arrived to universal surprise, others crept up on us step by step. The transistor radio came as a shock to people used to cumbersome 'wireless sets' plugged into the electric mains; the pocket calculator astonished a generation who knew calculating machines only as expensive office equipment; the digital watch and the portable phone were largely unexpected by people around the world.

Why the surprise? Obviously, for the people who made these products possible there was nothing surprising about them. Why the world found them surprising was because very few people were responsible for developing the technologies that produced them. Very few people knew what was about to happen.

Even though the electronics industry will become the world's largest industry within a decade, overtaking the car, steel and pharmaceuticals industries, very few people understand the driving force behind the electronics industry, which is the science of microelectronics.

Those few people are concentrated in relatively few areas of the world: California's Silicon Valley, the Texas plain, Arizona, Japan, Taiwan's Hsinchu City and South Korea. In Europe, the universities and the laboratories of the national telecommunications authorities, the various defense ministries and a few big companies provide a home for some more. If those areas were devastated by some disaster, the progress of the world's electronics industry would come to a shuddering halt.

The microelectronics priesthood resembles the medieval alchemists, in that the silicon priest's art and the alchemist's art are both 'black' in the sense that they are uncontrollable and unpredictable. Even nowadays, a multi-billion dollar, multi-national microelectronics company will find its production lines have gone haywire for no apparent reason.

Often it will be a fluke tweak to the process that does the trick. Just as disasters are remedied by a lucky break, so too many of the major advances have occurred almost by accident with the science trailing along some time later to explain it.

For instance, the invention which made possible the transistor radio, the portable, affordable version of the 'wireless', was the transistor. That was invented as a lucky by-product of research directed towards other ends and, even when it had been invented, very few thought of it as the basis for a portable radio.

The reason the transistor radio happened was because a material was found that could do the job previously done in wirelesses by glass bulbs. The material was first germanium, later silicon. It could take a faint electric signal – such as one that can be transmitted through the air – and boost it to an audible level.

In the old wireless sets that boosting was done by a glass bulb enclosing a vacuum – an invention of Thomas Edison. The bulbs were bulky and fragile and used lots of power, which is why wirelesses had to be big and heavy and plugged into the mains.

The silicon replacement for the glass bulb made it possible to use a small blob of material instead of a bulky glass shell enclosing a vacuum. The blob, named the 'transistor', could be very small, was apparently infinitely shrinkable, was rugged and used little power.

When they first appeared in 1953 (some five years after the invention of the transistor), the first transistor radios were greeted by the world's young with such enthusiasm that they became one of the largest selling electronics products of the 20th century.

The transistor was such a hit because it was cheap enough for young people to buy and portable enough for them to listen to it anywhere. It liberated the young by giving them control over their radio listening. Before the transistor radio came along, control over the expensive, immobile wireless set rested in the hands of their parents.

As the transistor radio became a familiar item in the world's shopping centers, so radio frequencies became dedicated to entertainment for young people, preparing the way for the youth culture of the 1960s, and the world's newspapers started filling up with complaining letters from those whose previously peaceful visits to parks and beaches were being ruined by the infernal but ubiquitous new noise-machines.

The impact of the transistor radio was so great because it came as a total surprise to most people. Whereas most technical inventions are flagged in advance and their improvement is a slow-change evolutionary process thereafter, the portable radio had not been foreseen.

There were two reasons for that. First, the scientific advance that made the portable radio possible, the invention of the transistor at AT&T's Bell Labs by Walter Brattain, John Bardeen and William Shockley in 1947, was a major Nobel Prize-winning scientific breakthrough in itself – unforeseen by everyone except a few scientists working in the field. Even Brattain admitted that it was a field in which he had 'an intuitive feel for what you could do, not a theoretical understanding'. Indeed, for seven months Bell Labs kept the invention a secret while it tried to find out precisely how and why the transistor worked – not wanting to be embarrassed by announcing an invention without understanding it.

The second reason for the surprise caused by the transistor radio was that hardly anyone saw the transistor as being suitable for producing portable radios. Two who did were the co-founders of Sony, Akio Morita and Masaru Ibuka. However when they asked Western Electric, AT&T's manufacturing arm, for a patent license on the transistor, Western Electric advised them to forget making transistors for radios and instead to use them to make hearing aids.

Fortunately for the world's youth, Morita and Ibuka persisted with their aim and produced the world's first portable radio. The story is a classic example of the relationship between the science of microelectronics and the electronics industry. The industry can't exist without the scientists, but the scientists often fail to see in their inventions the same potential recognized by the industrialists.

When great science coincides with great industrial vision, the result is what the industry calls a 'killer' product – one that surprises and delights the world, and which the world cannot resist buying.

No less of a surprise was the introduction of the calculator. Again the astonishment of the public was because the machine was completely unexpected. Whereas offices had large heavy calculating machines costing thousands of dollars, ordinary citizens checked their bank balances and tallied up their tax returns aided by nothing except mental arithmetic.

So when the first cheap, pocketable calculators appeared in the early 1970s they offered the prospect of putting an end to the mental toil which arithmetic involved for those of the population not blessed with an aptitude for figures – a good many.

Like the portable radio, cheap calculators came as a complete surprise because no one had anticipated them. They became possible because of a major microelectronics advance that occurred between the invention of the transistor in 1947 and the arrival of the first pocket calculators in the early 1970s – the invention of the chip.

Whereas the transistor radio used the ability of a transistor to accept a weak broadcast signal and boost it to audible levels, the calculator used quite another capability of the transistor.

The use which the calculator exploited was the transistor's ability to switch on and off quickly. By using its 'on' state to represent a '1' and its 'off' state to represent a '0' the transistor could be used to store information translated into the 'binary' language, which is made up solely of 1s and 0s.

Because early computers had used the vacuum-filled glass bulb to act as a '1'/'0' switch they occupied whole rooms and used enough electricity to power villages. By substituting a silicon blob for a glass bulb, massive size and energy reductions resulted.

The beauty of the blob was that it could be continually reduced in size while retaining its capabilities. That meant it was continually cheaper to produce and took up less room in electronic products, allowing them, in their turn, to become smaller.

However, this shrinking process was dramatically speeded up as the result of a major microelectronics breakthrough in 1958. That was the first successful fabrication of two transistors on the same piece of material – the world's first chip. Soon the number of transistors that could be put on a chip was to escalate.

During the 1960s it became possible to put several hundred transistors on one chip, and by the early 1970s it was possible to put several thousand transistors on the same-sized chip. Since a few thousand transistors are all it takes to make up the electronics of a calculator, by the early 1970s the first cheap pocket calculators were produced.

Like the transistor radio, cheap pocket calculators caused universal surprise and were a liberating force. Where the transistor radio had liberated the young by giving them personal freedom over their choice of broadcast entertainment, the calculator liberated the innumerate from the tyranny of arithmetic.

The subsequent history of the calculator – declining in cost to the point where it is found in a child's Christmas stocking – is a classic tale of how advances in microelectronics reduce cost. It also demonstrates a truth in the electronics business – that all products become mass-consumer items when their electronic content is reduced to a single chip.

That's because it is a rule of thumb in the microelectronics business that even the most complex chips eventually come to be sold for a few dollars. So when any product gets down to the level of needing one chip, it is on the way to becoming a Christmas stocking item.

Although the transistor radio and the calculator caused the biggest surprises, they are not the product that has shown the most stunning advance. That honor must be reserved for the computer, the product that has shrunk from a roomful to a lapful in 50 years, as shown in Fig. 1.1.

Maybe the relative lack of surprise caused by the computer's evolution is because the advance happened gradually and was well flagged in advance. By the time people were able to afford computers, in the 1980s, most people were well aware of their capabilities. So their adoption was more of an evolutionary affair than a surprise.

The microelectronics invention that made the computer affordable for all, or 'democratized' it, to use the expression of its inventor, Ted Hoff, was the microprocessor.

It happened as the result of a meeting in 1969 between the president of a Japanese company called Busicon, Mr Kojima, and the president of the American company Intel, Robert Noyce, who is co-holder of the patent on the first chip.

Kojima commissioned Intel to produce eight to twelve chips for a calculator. Hoff was the engineer entrusted with the commission, but instead of designing a bunch of chips to perform solely the functions of a calculator, Hoff produced an all-purpose set of chips that could be programmed to act as the electronic guts of many different products. That is the definition of a microprocessor, and Hoff's chip-set was the first implementation of the concept.

Because Hoff's microprocessor could be used in many different products and could therefore be sold to many different kinds of

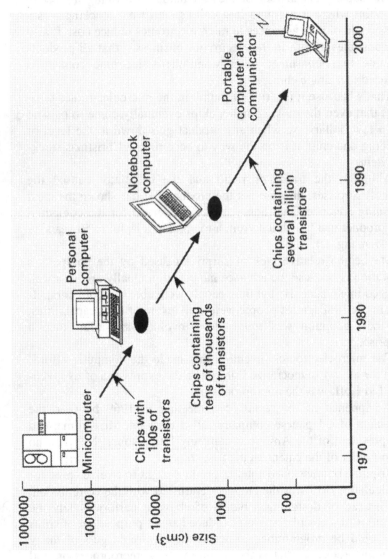

Fig. 1.1 How the shrinking transistor changed the computer.

manufacturers, it was something that could be churned out in high volumes and consequently cheaply.

Recognizing the potential, Intel bought the rights to the design back from Busicon and proceeded to develop a series of microprocessors which continues to this day. In the mid-1990s, Intel's microprocessors powered some 80% of the microprocessors in the world's personal computers.

That is not to suggest that a PC needs only a microprocessor. When, in 1981, IBM used an Intel microprocessor to make its first PC, the computer required over two hundred chips in addition to the microprocessor. Ten years on, simple PCs commonly use a dozen chips and the single-chip computer is not far away. That is when computers will be slipped into Christmas stockings.

The transistor radio, the pocket calculator and the portable computer are all very visible examples of the effect of the shrinking transistor, and most of us have bought one or another of them. However, the shrinking transistor has affected us all in less visible, but no less potent, ways.

Industries that used to employ massive unskilled or semi-skilled workforces have drastically reduced their headcounts as electronic machines get cheaper and more powerful and make a few people as productive as many. Wherever the phrase 'technology advance' is used in an electronics context, what it comes down to is the effect of the incredible shrinking transistor.

But what the electronics industry has achieved so far is peanuts compared with what it is about to do, for one simple reason: from the invention of the first chip in 1958 until today the number of transistors you can put on one chip has doubled every two years. Each doubling up doubles performance, while the cost remains almost constant.

Between 1968 and 1993 the cost of making one transistor on a chip fell from 2 cents to 0.0003 cents and the number of transistors you could put on a commercial chip went from 1000 to 16 million.

So, for over 30 years, the cost of producing an electronic product has halved every two to three years or, to put it a different way, the capabilities of electronic goods have doubled every two to three years without costing any more.

Why microelectronics now stands on the brink of an explosion in capability is because, whereas in the 1970s the doubling up related to transistors in their thousands and through the 1980s it

related to millions of transistors, in the 1990s it will be about
billions doubling up, and that represents enormous increases in
capability routinely being dumped on the doorsteps of anyone
making electronics products.

The question being asked is 'What will be done with all that
capability?'. No one really knows. The safe answer is to predict
that there will be more of the same: evolutionary improvement of
existing products and, every now and again, a major surprise.

However, that doesn't take into account one thing: with more
and more people using electronics products and having their own
ideas about what they want, and with more and more people
understanding the ever-declining cost base and ever-improving
technology base on which electronics goods are made, there are
going to be many more brains to tap for good ideas.

The overall effect should be a 'democratizing' one as more and
more people get to be able to use and afford powerful electronic
tools. Just as the transistor radio liberated the young from having to
listen to programs selected by their parents, and the pocket
calculator liberated non-mathematicians from the tyranny of arithmetic, so the new products will give more and more power to
individuals.

For instance, the televising of the Vietnam War – the first war
which communications satellites could bring directly into peoples'
homes – was brought to an end by the outrage of the American
people about what they saw. That was achieved by the electronic
technology of the 1970s.

Nowadays, video cameras, 'camcorders', are becoming affordable by many and when linked into cellular telephone networks
people will be able to transmit their pictures instantly to their television back home or to someone else's television. That is a powerful capability for the ordinary citizen.

If the official broadcasting organizations can stop a war by
broadcasting it, then the forthcoming ability of individual citizens
to show each other what is going on in the world will be a strong
force for truth. Word of mouth is a more credible medium than the
words of the commercial media organizations and very much more
credible than the words of governments.

Already the ability of satellites to broadcast over the boundaries
of countries is a strong encouragement to governmental honesty
and governmental commitment to looking after the interests of
their people. For if people can see on television that life is better

elsewhere, they will want to emigrate – and that should keep governments on their toes in providing a good life for their citizens.

For citizens of the relatively wealthy countries The Age of the Global Nomad will be at hand in the next decade. The vehicle for doing this is the wireless digital communications networks that are currently being set up around the world. Once they are in place, phone calls, faxes, and the sending of data or moving video pictures will be possible from anywhere to anywhere at a progressively decreasing cost.

At the heart of it is the shrinking transistor and its ability to store the 1s and 0s of the binary language. Since the binary language can store any form of information, whether it be spoken words, written words, photographs or video film, that means the transistor can store any form of information – pictures, video, voice, film, speech, script or music – and can communicate it down a wire or through the air.

Once all these different forms of media are reduced to the common form of 1s and 0s, you will only need one machine to handle the lot. And because the transistor on the chip roughly halves in cost every two years, so will the cost of storing information and communicating it using such a machine.

Twenty years ago people were being surprised by the first 'hole in the wall' banking systems. Nowadays a credit card can be 'swiped' on one side of the world and verified on the other in a matter of a few seconds. But these are elementary tricks compared with what will happen when global digital networks are in place.

Helping (some would say hindering) the emergence of digital networks are the telecommunications authorities in Europe, Japan and America, which are in the process of setting up telephone networks that can handle binary information. There are no technical problems to be solved in setting up these digital networks: it is merely a question of allotting wavebands and setting standards so that all the equipment made for using the networks can work together – 'compatibility' as the industry calls it.

When the Americans, Europeans and Japanese have got their digital networks up and running, then, assuming Washington, Brussels and Tokyo have the good sense to agree on compatible standards, it will be possible to send a fax from your laptop computer on a Californian beach which prints out on the fax machine in your office in Tokyo. Or transmit a video from your camcorder in Kyoto to your home in London.

Then there could really be some dramatic changes in lifestyle, with the removal of the pressures on people to congregate in cities. Already the professional workstations used by the designers of a vast variety of products, from aeroplanes to bridges, are becoming both cheap enough for individuals to buy, and portable enough for them to carry and use anywhere.

With digital networks allowing users of portable computers and workstations to access information anywhere in the world and with the addition of telephone, digital television, fax and video communications to the workstation/computer (features collectively called 'multimedia' by the industry) computers and televisions will become electronically indistinguishable.

So a single, all-purpose, combined computer/TV/communications tool will make it possible for people to live and work anywhere – speaking to colleagues, looking at office-based computer files and receiving broadcast news and entertainment. Such a thing might be selling for around $5000 in 1997 but, for certain, the shrinking transistor will ensure it gets cheaper all the time.

What the tool will do is provide the 'nomadic' lifestyle to all who do not need to inhabit an office for administrative, clerical or politicking purposes. It would be a boon to cities by relieving them of the congestion of commuters and a boon to deprived rural areas by providing them with a high-earning resident population.

The key to making it all happen is making the technology usable. Nowadays you can theoretically use a computer to send a message to another computer or to a fax machine, but when you are on the other side of the world, and unable to plug your computer into a telephone socket because all the sockets are hard-wired, you can't even get started.

With a global wireless digital network in place you will merely dial the number you want to send your messages to and press 'print' or 'send' to have your message printed out on a fax machine or stored in someone's computer. And another distinction that will go is between the fax machine and the printer. The same tool will do both jobs.

Indeed, there is much to be said for having your daily newspaper printed out in your own home on your own fax machine/printer rather than having huge wads of paper carried around the country. The contents of a daily newspaper could be digitized and broadcast to your home-based fax/printer in a few seconds, saving oceans of petrol in transport costs. If attached to a paper recycling machine, the printer could use the same paper over and over again.

The beauty of the digital network is that it will be possible to send large amounts of data quickly. So a full-length feature film could be transmitted to your computer/TV in a matter of seconds and then held in store while it is played over the next hour and a half or so.

Many people would say that all the uses of electronics so far mentioned are trivial compared with those that mimic the human senses – sight, smell, taste, touch, hearing and thinking. We humans are electrical in nature and so are the signals handled by chips. Already universities have produced silicon chips that can be implanted in the human body to replace damaged nerves to restore the flow of signals from the nervous system to the brain.

And, to an extent, the five senses and the brain can already be replicated in silicon. Limited machine vision and touch are available to industrial robots. Machine hearing is something on which a lot of research money is being spent around the world and a great deal of secrecy attaches to progress. Machines can recognize individual words, but are not so hot when it comes to sentences. None the less, the best people in the field think that machines capable of recognizing human speech will be on the market in the 1997–2000 time frame.

That opens a vast range of product possibilities from dictation machines that print out speech, to portable translators that speak out in a different language from the one spoken into them, and to a whole bunch of voice-operated machines, from telephones to cars to computers.

Smell and taste have had less research money than sight and hearing, probably because there are fewer commercial applications for them. However, the area that has attracted by far the largest amount of research money and effort is the area of artificial intelligence – finding a silicon-based equivalent to the carbon-based human brain.

Although silicon chips are already much better than human brains at storing information and infinitely better than human brains at performing mathematical calculations, they are none the less immeasurably worse than an insect's brain at tasks like recognizing a predator.

Computers, as we have seen, are based on the mathematical logic of 'true or false', '1 or 0', 'on or off'. But the real world isn't like that, seeing gray scales and infinite subtleties. 'In so far as the laws of mathematics are certain they do not relate to reality and in so far as they relate to reality they are uncertain', said Einstein.

So the chip-makers are looking at ways to model the human brain. In fact, they have been doing so for 40 years using software models, but only since about 1993 has the physical size of the transistor blob become small enough to make it possible to usefully copy the basic building block of a human brain, the neuron, in silicon.

Unlike the normal silicon chip, which contains a lot of storage space or processing power, the human neuron has small individual capability but lots of connections to other neurons, linking together with them in a web. Until recently, a big problem was that no one could find out how to get the whole web to recognize and store information picked up by an individual neuron allowing the whole web to take account of it and update its collective information. In the last five years a technique called 'back propagation' has made that possible.

The result is a technique for making self-learning machines that could be trained to recognize patterns. The patterns could be pictures, animals, human faces or sound such as music or speech. Once that has been achieved, the superhuman mathematical capabilities of computers could be married to an ability to recognize the same sensory inputs as humans receive. Instead of thinking like computers, computers would start to think like humans.

One of the great advantages of computers that think like humans is that they will be able to help us in complicated tasks like curing ourselves when we're sick or giving legal advice when we're in trouble. At the moment, computers are pretty bad at following the human thought patterns that would allow them to pick up the knowledge of a doctor or lawyer.

Experts such as these use 'if–then' types of reasoning to come to their conclusions, e.g. 'if the patient has a temperature then she may have a fever'. However, there could be a lot of other possibilities, and it is the job of human experts to process a succession of 'if–then' inferences until they arrive at the correct one.

Chips that are suitable for processing a great deal of such inferences very quickly are now on the market. Currently they provide fairly simple control functions for products such as 'if getting hotter then cool it'. They are called 'fuzzy logic' chips because they deal with gradations or gray scales of logic rather than the true/false logic of mathematics.

As these capabilities advance to make computers think along the lines of human beings, so it will become possible for us to find computers as easy to deal with as another human and as useful as a skilled human.

The great breakthrough these techniques of fuzzy logic and neural computers offer is the ability to talk to a computer in ordinary language and have it talk back in the same language. The practical benefit of that will be that you will be able to have a doctor when you're sick, a motor mechanic when the car goes wrong, a lawyer when you're in trouble and an accountant when you're sorting out your tax.

In an age when expert advice is becoming increasingly expensive – to the point where many in the West cannot afford medical and legal advisers – machines that provide this advice inexpensively will come as a boon to individuals and a relief to governments groaning under the cost of public health care programs.

Some might say they would never trust a machine-doctor. But they probably don't think twice about flying in aeroplanes, which are regularly landed by totally automatic electronic equipment. Just as the aircrew are nowadays there to deal with emergencies and to reassure the passengers, in the future a comfortable, friendly sort of person may act as a front for a mechanical doctor.

And to many people it might be a comfort to have a mechanical 'Mr Know-All' in the corner of the living room who isn't a bore and who only talks when spoken to.

For the experts themselves the future is also hopeful. Already much of the tedious detail of their jobs has been eliminated by computer programs. Programs, for instance, that calculate stresses for architects and builders can save an expert the trouble of working out how thick a beam or cable needs to be. The result of having the donkey-work done for them should free the designers of everything from bridges to automobiles to design at higher and higher levels of abstraction.

The logical result of that could be that anyone can design anything using a computer. In some areas that is happening already – look at the creativity behind a Swatch wrist watch – where the most important differentiating factor of the product is pure creativity and no technical knowledge of watch-making is required by the designer.

In other areas, for instance cars, design-by-computer currently tends to make them all look the same because each manufacturer is feeding in the same requirements. However once the engineering is common to all – like the mechanics of a watch – then designers will be employed simply for their creativity.

The ability of electronic progress to allow people to work at home should also make it as enjoyable for them to be entertained at home as it is to travel for entertainment.

Crude early versions of 'Virtual Reality' systems are with us now, where you can put on a helmet and a suit and have the experience of being somewhere else. A video scene is played in the helmet and as the head is moved more of the scene unfolds – as in real life – while the suit provides appropriate sensations of touch. For instance, if you reach out to touch a tree displayed on the video you will feel its bark on your hands.

Such a system allows for the 'armchair holiday': the ability to see and feel a new place without going to it. The whole 'holiday' could be digitized and squirted down a phone line to your computer. It could make the airline industry obsolete.

Already, 'Virtual Reality' systems are being sold around the world to companies that design very complicated products, such as in the aerospace industry. Virtual Reality allows such designers to work in 3D and to model their work as they do it.

However, the ultimate challenge for microelectronics will be to find ways of directly connecting the electrical signals of chips to the electrical signals of the nervous system and the brain. Various research programs around the world are looking at exactly this problem.

Recently, Fujitsu and Hokkaido University in Japan announced that they had identified the electrical signals in the brain associated with communications and were able to recognize, and differentiate between, the brainwaves generated when thinking different thoughts. It is, of course, a very long way from the 'thought input' computer, but the possibility is there.

The further implications of being able to communicate directly between the human brain and a silicon brain are immense. For instance, it might be possible to 'download' a brain into a silicon equivalent. If that was then connected to your brain you could experience what it is like to think someone else's thoughts.

Furthermore, it would be possible to transfer the contents of a person's brain before death and so keep their knowledge and experience available to succeeding generations. If you consider that to be fanciful, remember that it would have seemed fanciful to people of the 18th century that we would one day see people walk and talk after they are dead – which we now do via film and video.

The encouraging thing about the incredible shrinking transistor, which has made all these things possible, is that its potential for further dramatic advance seems limitless. In 1993, a combined research team from the Hitachi Cambridge Laboratory and the Microelectronics Centre of the Cavendish Laboratory at Cambridge University established that it would be possible to make a 0/1 switch using a single electron.

That would allow the equivalent of a 1 million million (1 000 000 000 000) transistor chip to be fabricated. At the current pace of technology evolution it would take some 30 years to get there. But before that level is reached, it looks as though it will be possible to reproduce in silicon the workings of the brain and the senses.

2

The pioneers

It all started with the tube (or thermionic valve). Before the tube, electricity was used for light, heat and for powering electric motors. After the tube, electricity was used for broadcasting, telecommunications and computing.

That transition represented the passage from the electrical age to the electronic age. The former provided an alternative to human muscle, the latter provides artificial means of performing other human functions such as sight, talk, hearing, memory and thought.

It is no mystery why it should be so. Humans are electrical in nature, with brains using electrons to store information. So does a computer. Limbs are controlled by streams of message-carrying electrons sent by the brain, just as telephones pass message-carrying electrons along wires.

The presiding genius over the passage from the electrical age to the electronics age was Thomas Edison. Edison discovered the operating principle of the tube and it became the basic building block of all electronic equipment.

The tube did two things. It boosted the strength of an electrical signal and it could turn an electrical signal on and off quickly.

The booster capability is used in electronic equipment to take an electrical signal which is carried through the air or along a wire and boost it into one that can be heard by the human ear or seen by the human eye.

The on-off switching ability of the tube is useful in electronic equipment because it allows the tube to be in one of two states – On or Off – so providing a tool to represent the 1 or 0 of binary language.

Almost anything can be translated into binary language, e.g. speech, writing, music, still pictures, broadcast TV pictures and video pictures. So a tool like the tube, which can store and transmit binary language, becomes very useful.

Besides providing the basic building block of the electronic age, Edison provided something else – a role model. Largely self-taught, Edison never went to college and had an anarchic, iconoclastic temperament which was to be shared by many of the most influential innovators in the succeeding evolution of microelectronics.

Edison mocked the scientific establishment ('long-haired' was a favourite term of disapproval), and he pursued a method of working which reversed the conventional scientific approach of first doing pure research and then using its results for applied research.

Instead, Edison performed practical experiments and his discoveries came by chance. 'At the time I experimented on the incandescent lamp [the light bulb] I did not understand Ohm's Law', said Edison, inventor of the light bulb.

He bridged the all-important gap between science and the market, always inventing for manufacturing rather than inventing for the sake of technological advance. He worked side by side with a factory operation, inventing products to feed the production lines.

For Edison, invention had to be related to a product that people would buy. A hundred years later his spiritual heirs in Silicon Valley pursue much the same formula in much the same spirit.

An example of his style was his 'invention factory', where, he boasted, he would invent something minor every ten days and perform a 'Big Trick' every six months. The invention factory's staff found their boss an enigma, veering from the charmingly entertaining to the tyrannical, but they delivered the goods – in 1877 the invention factory's two big tricks were the gramophone and the microphone.

Market recognition for those two inventions came a good deal earlier than for Edison's great building block – the tube. The first uses for the tube came in radio communications and broadcasting. The tube's use in telecommunications along telephone wires and in electronic computing came later.

By taking a signal which could be transmitted through the air and then boosting it to make it audible, the tube became the building block for the radio. In 1895, Sir Ernest Rutherford transmitted radio signals over half a mile; in 1901 Guglielmo Marconi demonstrated transatlantic radio. Radios were then seen as useful point-to-point communications tools, especially for ships at sea.

Then, on Christmas Eve 1906, the first broadcast was made in Massachusetts, comprising a couple of tunes, a poem and a short talk picked up mainly by ships' wireless operators up to a few hundred miles away.

In 1919 the first broadcasting station was set up at Chelmsford in the UK, which sent out two half-hour programs a day. A particularly popular one was by the opera star Nellie Melba, but the station was closed down by the British Post Office.

The British Post Office's reaction to new technology had often been quirky. The Postmaster General commented on Bell's invention of the telephone: 'While the Americans might need such a daft thing, Britain has plenty of small boys to run around with messages'.

The Post Office's official attitude to the new phenomenon of broadcasting was to require that every transmitter and every radio be licensed. By the end of 1921, it had licensed 4000 radios for receiving and 150 broadcast stations for transmitters. The following year the BBC was set up and the Marconi Company was granted a license to broadcast for 15 minutes a week.

The Americans were more laid back about broadcasting and the world's first commercial radio station, KDKA of Pittsburgh, opened for business on 2 November 1920, with news of the voting returns in the presidential election, won by Warren Harding. A year later there were eight US broadcasting stations and by 1922 there were 564.

Elsewhere the story was the same. The first Canadian broadcast was in 1920; in 1921, Australia, New Zealand and Denmark started; in 1922, France and Russia followed; in 1923, Belgium, Czechoslovakia, Germany and Spain; in 1924, Finland and Italy; in 1925, Norway, Poland, Mexico and Japan; in 1926, India.

Sales of radios rocketed. TV broadcasts soon followed. In 1935, TV programs were first broadcast in Germany; TV began in the UK in 1936 and in America in 1941. The Second World War interrupted the proliferation of TV but, after 1945, it was rapid. In 1940, there had been 10 000 TVs in the USA; in 1949 there were 1 million; in 1951, 10 million; in 1959, 50 million.

In 1962, using satellite-borne tubes in space to receive radio signals from Earth, to boost them and then to retransmit them back to another spot on Earth, the first extraterrestrial broadcast was demonstrated by AT&T. In 1964, the Tokyo Olympic Games was broadcast via satellite to America.

So, although it had been expected that the tube's main use in communications would be in person-to-person communication, it actually achieved its first widespread use in the new field of broadcasting. That pattern was to be repeated with the transistor, and later the chip. In both cases, a device thought to be useful for one purpose actually made possible totally new activities.

It was not until 1915 that the tube started to be used in wired telecommunications as an intermediate booster for signals along telephone wires. Without a boost the signals would fade over distance; with a boost, the signals could reach longer distances.

That was suitable for land-lines, but underwater trans-oceanic lines were a different matter. That was because tube lifetimes were short, and whereas it's a simple matter to replace a dud tube in a land-line, it's impossible when the telephone line is on the bottom of the Atlantic.

So trans-oceanic telephoning had to be done by radio. Before 1956, a transatlantic phone call started by going down a wire, crossed the Atlantic by radio waves and ended its transmission along a wire. The first radio transmission of a voice across the Atlantic had taken place in 1919 and the first commercial transatlantic telephone service opened in 1926.

However, by the mid-1950s tube technology became sufficiently advanced to allow tubes with a guaranteed 20 year lifetime. That was considered to be sufficient to make it worthwhile using them to boost signals in trans-oceanic cables and the first transatlantic telephone cable was laid in 1956.

After pioneering the broadcasting industry and revolutionizing the telephone industry the third major impact the tube had was in the computer industry.

It was the Second World War that acted as the catalyst. In England the group of scientists at Bletchley Park decoding messages sent by the German Enigma encoding machine could not always perform their task quickly enough to pass the information on to the politicians and generals in time for effective action to be taken. At that time they used a mechanical decoder.

A Post Office engineer called Tommy Flowers suggested that the Bletchley Park group should build an electronic decoder using tubes. The suggestion was rejected and Flowers set out to build it himself.

In 11 months he and a team of helpers completed the 1000 tube machine. Colossus, as the machine became known, not only worked, it was dramatically faster than the mechanical decoder. Coming into service in December 1943, it made an incalculably significant contribution towards winning the war.

In America, the Ballistics Research Laboratory was looking for a quick way to calculate the trajectory of shells under differing environmental conditions to give gunnery officers the appropriate settings for their artillery. As at Bletchley Park, they found the problem they faced was insufficient speed in working out the calculations in time.

The laboratory set up a research unit at the University of Pennsylvania and engaged one of the university's professors, John Mauchly, to build a tube-based calculating device. Mauchly and a colleague, Presper Eckert, built the 18 000 tube, 30 ton 100 ft × 8 ft × 3 ft ENIAC machine – costing half a million dollars and completed (in November 1945) too late to be used in the war.

Colossus and ENIAC were the first tube-based computing machines and so the first electronic computers. They spawned the mainframe computer industry, grandparented the mini-computer business and great-grandparented the PC, which, in the mid-1990s, reached 50 million-plus unit quantities, increasing with every year.

With the computer industry adopting the tube, the tube had become the basic building block for electronic equipment. It had made possible radio and TV broadcasting, long-distance telephony, and the first useful computers. The tube also became the standard screen for TVs, computers and electronic instruments.

As the building block of electronic equipment, the tube's development became the main force for the development of the electronics industry. But the tube was a pain to work with. Being made like a light bulb – a glass dome enclosing a vacuum – it was very fragile. ENIAC had to have a tube replaced every few minutes.

Tubes also used a lot of electricity and created a lot of heat. Miniaturizing the tube and making it more rugged and less power hungry looked like being the way forward for the technological progress of the electronics industry.

Had it been so, the development of the electronics industry would have been a much more staid and slow-moving process than it turned out to be. Fortunately, on 23 December 1947, Bell Labs, R&D arm of the US telephone operator AT&T, gave the world a great Christmas present – the transistor.

As with the tube, the initial discovery was accidental. Bell kept it secret for seven months while scientists worked on figuring out how it worked, thinking up a name for it and filing patents. On 30 June 1948 the transistor was unveiled to an unenthusiastic world, with next day's *New York Times* giving it 4½ inches on page 46, referring to it as: 'A device called a transistor which has several applications in radio where a vacuum tube ordinarily is employed'.

Elsewhere the invention had more of an impact. In Tokyo a young sub-lieutenant later de-mobbed from the Japanese Imperial Navy took keen interest. He was Akio Morita, founder of Sony. And in Grinnell, Iowa, the physics teacher at the local college obtained two of the first transistors ever made from his old university classmate, the transistor's co-inventor John Bardeen, and showed them to his fascinated star pupil, Robert Noyce, the future founder of Fairchild Semiconductor and Intel.

Just as the tube took years to achieve its potential, so did the transistor. Its first use was in hearing aids, where the miniaturization and lower power requirement provided obvious advantages, which outweighed its extra cost over the tube. Moreover, Bell Labs waived royalty payments for transistors used in hearing aids as a gesture towards the work of Alexander Graham Bell in helping deaf people.

However, despite the boast of the president of one of the early transistor manufacturers, Germanium Products, that 'We expect to chase the vacuum tube price to hell and gone', germanium did not withstand high temperatures, so limiting its applications. This in turn meant that it was not likely to be produced in high volume, which alone could have brought down cost and price.

Everybody thought that it would be better if transistors could be made out of the more robust and available material silicon but, as speaker after speaker at the 1954 meeting of the Institute of Radio Engineers bemoaned, the technical difficulties in making transistors out of silicon meant that it would be many years before a silicon transistor could be made. Then up got Gordon Teal of Texas Instruments – a geological research company that had diversified into transistor production only the previous year – and gave a paper describing how they had made a silicon transistor.

Immediately the industry perked up. In 1951 there had been only four companies making transistors. In 1952 there were eight, and in 1953, 15. By 1956, that had risen to 26. Also, the US military, attracted by the new high-temperature capability of the silicon transistor, started the development funds flowing, so triggering the all-important cost-reduction process.

In 1952 the US Air Force, naturally the branch of the military that was the most concerned with reducing the weight and size of electronic equipment, had estimated that 40% of its electronics could be handled by transistors saving 20% in size, 25% in weight and with 40% fewer failures. When the silicon transistor came along, the military began a $15 million transistor development project, starting in 1956.

Whether it went to the best people was more doubtful. By 1959 the tube manufacturers making transistors as a sideshow were getting 80% of the US government's development money, while the new companies that had set up to make only transistors were getting 20%. This was happening while the transistor-only companies had over 60% of the market.

It was a pattern that was to recur in the microelectronics industry, with the big companies using their muscle to cream off government money, while smaller, newer companies – with less government help – were none the less taking the market away from the established players.

In the 1950s, the established tube manufacturing companies were Western Electric, General Electric, RCA, Raytheon, Sylvania, Philco, CBS (Columbia Broadcasting System), Tung-Sol and Westinghouse. The earliest companies to get into transistor-making without having been tube-makers were Hughes, Transitron, Germanium Products (which started up just to make transistors for hearing aids), Texas Instruments, Clevite and Motorola.

By 1955 the newcomers had overtaken the traditional tube-makers in transistor manufacturing. The 1955 top ten transistor-making companies were Hughes, Transitron, Philco, Sylvania, Texas, General Electric, RCA, Westinghouse, Motorola and Clevite. By 1957, the new companies had grabbed over 60% of the transistor market, a pattern that was to repeat itself over and again in the microelectronics business as each new change in the basic technology brought to the fore a new set of players in the industry. By 1975 all save three of the 1955 top ten had dropped out. The survivors were Texas, Motorola and RCA.

It was understandable why the tube manufacturers gradually slipped back in the race for transistor markets. In 1957, transistor production was a relatively small business – one twentieth the size of tube production. Naturally the tube companies concentrated more on their established business of tube production rather than the new, risky field of transistor production. By the same token, the fast-moving pace of technological advance meant that the established companies had no technical advantage – the newcomers were on the same technological level as the old companies.

However, whereas the tube market peaked in 1955 (by volume) and in 1957 (by value), the transistor market just kept on growing. The commercial applications were hearing aids and the transistor radios from Sony in Japan and from the collaboration of IDEA Corporation and Texas Instruments, which produced the $49 'Regency' transistor radio of 1954. However, in the late 1950s, over half the American market for transistors came from the military.

When the transistor was applied to the computer market it started a revolution. In 1955, IBM put on the market a computer which replaced 1250 tubes with 2200 transistors, reducing size, eliminating the need for cooling, and reducing the power usage by 95%.

The computer, called the 650, pioneered mass-manufacturing in the computer industry. In 1955 there were only 250 computers in the USA, 150 of which had been sold that year. In 1957, IBM took orders for 1000 of the low-priced 650s aimed at business users. The 650 was the computer industry's Model T Ford.

Ten years after the 650's introduction its effect had so stimulated the computer industry that 7400 computers were sold in the USA in 1965. That took the installed base of US computers to 31 000. By then computers accounted for a sixth of all the transistors produced.

The 1950s were good times for the transistor-makers. The seven-year-old company Transitron was valued at $285 million in 1959 and Texas Instruments' shares went up from $5 to $191 in the seven years between 1952 (the year before it got into transistor manufacturing) and 1959. That started off another revolution in the microelectronics industry, with engineers realizing that by starting their own companies they could get rich pretty quick.

If the main asset required to start a transistor company was the knowledge inside the heads of the transistor engineers (and it was), and if the rapid changes in the technology meant that the established companies had no technical advantage (which they didn't), then a bunch of transistor engineers could compete with the best companies in the industry and beat them (which they did).

For instance, Fairchild Semiconductor had been founded in 1957 and in 1960 was the eighth largest transistor company in the world. Five years later it was world No. 3, and by 1968 it was the world's largest transistor manufacturer. Meanwhile, groups of engineers had split off from Fairchild: in 1959 to found Rheem; in 1961 to start Amelco: and in 1963 to set up Signetics. One electronics genius, Jean Hoerni, was on the founding team of four companies in ten years: Fairchild, Amelco, Intersil and Union Carbide.

These companies were all hungry. They all wanted to be bigger and better than the rest, and the best way to do that was to crank up the volumes of transistors, which reduced costs, which allowed them to drop prices to beat competitors. A rule of thumb calculation by Texas Instruments at the time was that every time you doubled production volume you reduced price to 73% of the previous level.

That phenomenon remains the driving force behind the constant pervasion of microelectronics into new areas and new kinds of products. Every time the price went down, more new users of transistors came along having decided they could afford to use transistors in their own products.

The new users increased the overall demand for transistors, so obliging the manufacturers to turn up the volumes, so bringing down the prices some more, so further encouraging new users and so on and so on. That is the microelectronics industry's 'virtuous circle'.

In addition to the higher and higher volumes, another factor was helping to increase volumes while reducing cost. At the time there were rapid 'yield' improvements brought about by technical advances. Yield means the proportion of working transistors to dud transistors coming off the production lines. Yields had been dramatically increased by process improvements made at Bell Labs and General Electric and passed on to the industry at a second Bell Symposium in 1956.

These improvements made possible a more controllable kind of transistor, called 'mesa'. Immediately yields went up from a typical 10–20% to over 60%, vastly increasing production output and decreasing prices by the same proportion. That's because it cost as much to make a dud transistor as it did to make a good one.

The collective effect of all this effort throughout the 1950s to produce in high volumes was to bomb prices out of sight. Companies tended to sell transistors for whatever they could get for them. Adding to the intense competition was the first appearance on the US market of Japanese transistors in 1959.

The result was that the average price for a germanium transistor dropped from \$1.85 in 1957 to 50 cents in 1965. And a silicon transistor went from \$17 to 86 cents in the same time frame.

The effect on the computer industry of the improving cost/performance ratio of transistors was dramatic. In 1957, two years after IBM had introduced its 'Model T', a new company formed by Ken Olsen, called Digital Equipment Corporation (DEC), brought out a computer in 1959 which it called a 'minicomputer'.

DEC's minicomputer was ten times smaller and, at \$125 000, ten times cheaper than comparably performing computers (called 'mainframes'). Three years later a new, improved machine came out – it cost \$27 000. Three years after that, a better computer was launched – the cost was \$18 000. The Great Computer Boom was under way.

And feeding back into the transistor industry, the computer boom was expected to ensure that the Great Transistor Boom would carry on forever, with prices declining, volumes escalating, profits rising, new companies proliferating, technology constantly improving and fortunes continuing to be made.

As was to happen many times during the subsequent history of the microelectronics industry, those heady, hopeful expectations were dashed. This time it was by the slump of 1961–64. Throughout those years the industry had no growth in the value of its output. In 1964 the total value of the output of the transistor-makers was still the same as the industry's 1960 output, but, during the same four years, the number of transistors made by the industry actually tripled! The effect on prices can be imagined.

Many factors were blamed: there was a general economic slump during those years, the Japanese had entered the market in a big way, and the military share of the market had started to decline after 1961. However, the real reason was that the transistor industry had suicided through over-supply. There were too many companies making too many transistors and selling them for too little money.

There was another underlying industry problem that was spelling death to the transistor – wires. As electronic products became more complicated they needed more transistors, and more transistors meant more boards (onto which transistors were soldered), and more boards meant more wires to connect them together, and the more wires there were the more the chance of a connection breaking and stalling the whole system.

So the transistor industry needed a new product to get its prices back up to a sensible level and the electronic equipment industry needed something to get rid of wires. Invention, in this case, was the daughter of necessity. The invention which solved the problem was the chip.

It was nearly an English invention. In 1952, Geoffrey Dummer of the UK's Royal Signals and Radar Establishment told a Washington audience 'It seems possible to envisage electronic equipment in a solid block without connecting wires', and in 1957 he showed a non-working model of what he meant at an exhibition in the UK – it was the first model of a chip. Dummer couldn't get the backing to build a working model.

However, in 1958 a young engineer at Texas Instruments, working through the summer break because he had not been there long enough to qualify for a holiday, hit on the idea of making a complete electrical circuit in which all the different components – transistors, resistors, capacitors, insulators, rectifiers, diodes – were made out of the same piece of material.

It would have seemed an odd idea to most engineers because each of these components was normally made out of a different material – naturally the material that was able to deliver the best performance for that particular component. To make components out of a material that was not best suited to the component's intended function was against an engineer's natural mind-set.

The Texas chip contained transistors, resistors and diodes all made out of the same material, germanium.

In California, an engineer at Fairchild was working on the same idea of making all the components from the same material. Instead of wiring up the components on the chip, he evaporated aluminium connections onto the material, so fusing the connections onto the silicon and making what was essentially a solid block.

The process, called the planar process, had been invented by a Fairchild colleague, Jean Hoerni, one of the eight who had split away from doing research at William Shockley's semiconductor company. One reason for the split was because they wanted to pursue more practical goals than Shockley's. Whereas Shockley's goals were the goals of the scientist – focusing on the development of the transistor itself, the goals of the eight defectors were the goals of the technologist – how can microelectronics be mass-manufactured cheaply?

The Texas Instruments chip was completed and tested in October 1958; Fairchild's chip, using the planar process, arrived six months later. After battling for years about who had the right to claim the patent on the chip, it was agreed that it should be shared between the two engineers involved: Jack Kilby of Texas Instruments and Bob Noyce of Fairchild.

The transistor had been a scientist's invention – the result of pure research; the chip was a technologist's invention – the answer to a real-life problem. As Kilby himself modestly commented: 'It contributed very little to scientific thought'.

A contemporary assessment of the chip's significance summed it up: that 1000 tubes could fit into a cubic foot in 1950; in 1956, 10 000 transistors could fit in a cubic foot; and with the chip in 1959, a million components could fit in a cubic foot.

As with the transistor, the first commercial application was in hearing aids, where miniaturization and low power were of much greater importance than cost. That was important, because, when the first chips came on the market in 1961, the initial reaction was that they were terribly expensive – around $120 – much more than the cost of using individual components and assembling them by hand.

The American government's Moon and missile programs saved the day for the chip, by providing federal funds to kick-start the fledgling technology. The Moon rockets had to have an on-board computer to work out such complex problems as when to fire the motors to take them out of Earth orbit to put them on-course for the Moon and when to fire the motors to leave the Moon's gravity field to get them on-course for Earth. Such decisions could only be taken once and, if wrong, meant disaster.

There was no possible way of making a computer small enough except by using chips. Chips permitted the building of a computer for Moon rockets which was two feet long, one foot wide and six inches thick, weighing 54 lbs.

The other great military requirement for chips was in long-range missiles, where miniature computers were essential if the missiles were to find their targets. From carrying in their computers maps of the terrain over which missiles passed so that they could recognize their route, modern missiles have advanced to carrying complete 3D mock-ups of city centres so that they can find a particular building.

In the Gulf War, for instance, Cruise missiles were programmed with complete 'Virtual Reality' 3D models of downtown Baghdad so that they could guide themselves along streets, make turns at street corners and recognize their target buildings.

Back in the 1960s, the main commercial application for the chip was the computer industry. IBM first used the chip in 1964 in the computer that gave IBM its later dominance over the computer industry – the IBM System 360. The 360 was, at the time, the largest industrial venture in US corporate history, costing $5 billion, and it bet the company's future on the idea that computer users wanted a range of computers that could all work together and that all worked with the same software.

The System 360 set the standard, others followed, and within four years of its launch the number of computers being sold annually in America had risen from 7500 in 1964 to 147 000 in 1968, increasing the total number of computers in the USA, during the same four years, from 24 000 to 694 000.

With demand from the Moon and missile programs and with increasing use in the computer industry, the chip took off. In 1963 4.5 million chips were manufactured in the USA. Eight years later over 600 million were made and the old rule of the tube and transistor industries began to take effect – volume production meant lower prices. From an average US price per chip of $50 in 1962, ten years on the 1972 price was $1.

In the 1970s, the chip market boomed. The 1969 worldwide market for chips was worth $2.2 billion; by 1976, it had more than doubled, but the application that was to be its biggest customer had hardly been invented. Meanwhile, the ability of the chip to keep doubling its capability every two to three years while halving its average cost had effects which astonished an unsuspecting world.

The first of these effects occurred in 1973, when, to the disgust of the entire Swiss nation, electronic watches called 'digital watches' appeared. They were made possible by a combination of small screens (called light-emitting diodes – LEDs) which showed the time in numbers, and chips capable of storing the programs to change the numbers showing the time.

In 1973 these electronic watches cost $250. Two years later, they cost $150. Everyone piled into making watch chips and LEDs, and the price collapsed. At Intel, which bought a watch company in 1972 and sold it in 1978, Gordon Moore remembers: 'By the time we got out of the watch business the buttons on the side of the watch cost more than the chip'.

The second unexpected effect of the increasing power and declining cost of the chip happened to the calculator industry. In 1971, a basic four function (add, subtract, multiply, divide) electronic calculator cost around $100 in the USA. In 1972, it cost $50. By 1975 it had dropped to $20. Two years later it was $14 and by 1978 the cost was $11.

One of the companies nearly ruined by the plummeting price was a small calculator company called MITS, located next door to a massage parlour in New Mexico. In 1975 the owner Ed Roberts, faced with bankruptcy, took a final desperate gamble and decided to make and market a computer.

In a month or two he had put together a computer based on Intel's 8080 microprocessor and got a 19-year-old called Bill Gates to write the programming software. Roberts called the computer the 'Altair' and advertised it for sale at $500. A month later MITS had $250 000 in the bank from checks sent in with orders and a new industry had been born – the personal computer industry.

Later the same year, 22 Altair owners in San Francisco started up a club for themselves called the Homebrew Club. Two members were students called Steve Jobs and Stephen Wozniak. To show off something to the club, Wozniak put together some off-the-shelf chips, stuck them in a computer board and took it along to a club meeting. Jobs named it 'Apple I'.

After the Altair and the Apple I, the third leg to the personal computer revolution was the IBM PC. The Altair seeded the Homebrew Club which seeded the Apple I; what seeded the IBM PC was the fabulous success of the Apple I's successor, Apple II.

The Apple II was first sold in 1978 and made $700 000 worth of sales that year. The following year, sales were $7 million, and the year after $48 million. In 1980, sales doubled again and the Apple company went public, giving Jobs and Wozniak $100 million each.

IBM couldn't ignore that. In July 1980, it set up a project to get into the personal computer business within a year. Because of the time constraint, the project team scrapped the usual IBM practice of making every major component in the computer themselves and decided to build their computer from standard components that anyone could buy.

For the microprocessor IBM chose the Intel 8088. For the operating software IBM turned, as Ed Roberts of MITS had turned, to Bill Gates. Gates bought a system called 'the quick and dirty operating system' from a company called Seattle Computer Products. Realizing that 'quick and dirty' was hardly likely to appeal to his prestigious client, Gates renamed it 'disk operating system' – the nowadays ubiquitous DOS – before licensing it to IBM. The one year target slipped by a month or two, but, in late 1981, the IBM PC was launched.

The personal computer has become the greatest user of microelectronics and, in return, the advance of microelectronics has changed the personal computer. Ten years on from the desktop 1981 IBM PC, which contained over 200 chips, the microelectronics industry can deliver the same performance on six chips and the computer industry can wrap those six chips in a plastic case with a screen and a keyboard which will fit in your inside coat pocket.

How did that happen? The answer is, simply, the incredible shrinking transistor. By making the transistors on each chip smaller, each chip could contain more transistors and with more transistors the chip could deliver more performance.

For instance, each memory chip in the IBM PC of 1981 could hold 64 000 bits of information – enough for a three-page letter. Ten years later, in 1991, a memory chip could hold four million bits of information – enough to store 180 typewritten pages (Fig. 2.1).

The 1981 memory chip contained 64 000 transistors; the 1991 chip contained four million transistors. Clearly, the only way that trick can be pulled off is to make the transistors smaller. In the 1981 chip, the transistors measured three millionths of a meter across; in the 1991 chip the transistors measured only eight tenths of a millionth of a meter across.

That is why six chips in 1991 can do what 200 chips did in 1981. In the computer business, this increasing power has worked in two ways: first, to make very small, portable computers with the same power as early ones; second, to make more and more powerful desktop computers that can run very big software packages, which make the computer both easier to use and capable of performing more and more difficult tasks.

Higher performance is not just a matter of better memory chips. It's also a matter of working faster. Clearly, the faster a computer works, the more work it can do in a given time. The chip that defines how fast the computer works is the microprocessor chip, and between 1981 and 1991 the speed of microprocessors increased 100 times.

The reason for that is the incredible shrinking transistor again. The smaller the transistors are, the closer they are packed in together. The closer together they are means the less distance the

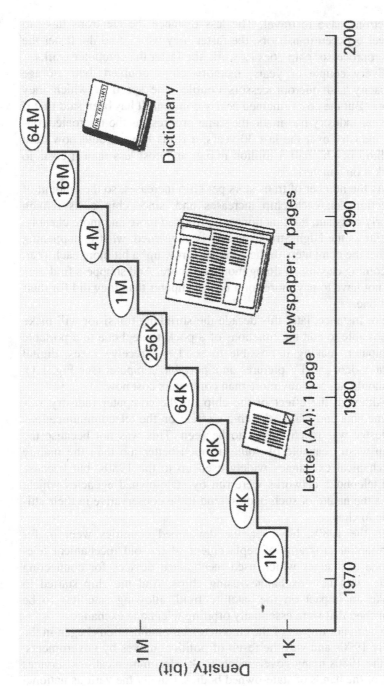

Fig. 2.1 What you can store on one chip 1970–96.

electrons have to travel. The less distance the electrons have to travel between transistors, the faster they work. And the faster the microprocessor chip does its work, the faster the computer works.

Every couple of years, memory chips doubled their storage capacity and microprocessors doubled the speed at which they work. But the cost remained nearly constant. It has been said that if the car industry had made the same progress as the microelectronics industry over the last 30 years, a Rolls-Royce could now do a million m.p.h., half a million m.p.g. and cost less than it costs to park it on a meter!

As the number of transistors per chip increases, so the amount of functions on each chip increases and, since chip costs remain nearly constant, the functions are provided more and more cheaply. So far as the High Street shopper is concerned, what is happening is that the computers in the window 'grow up' a bit more each year, becoming capable of doing more and more. And shoppers find they do not have to pay more for a new computer than they did for their last one.

For instance, later this decade the shrinking transistor will make it possible to add the functions of a pocket telephone to a portable computer, making it possible to send and receive faxes, digital data, video, and TV pictures on a portable computer (see Fig. 1.1). It should not cost any more than computers cost now.

Although the effect of the chip on the computer industry was immediate and major, the chip's effect on the telecommunications industry was much slower to be seen. This was not because the chip wasn't capable of doing a much better job than the mainly mechanical exchanges widely used up to the 1980s, but because the telephone networks were run by state-owned agencies which, as is the nature of such bodies, tend to be conservative in their attitude to change.

In the 1980s, however, the developed countries went in for substantial programs of replacement of the old mechanical telephone exchanges which used mechanical devices for connecting calls, with new exchanges using chips. And the chip started to make an impact on the satellite field, allowing satellites to be launched that were essentially orbiting telephone exchanges.

The major impact of the chip in telecommunications began in the early 1990s and was the result of political moves by governments. In the 1980s many governments took telecommunications services out of the hands of state-owned bodies, usually the various national

Post Offices, and let in commercial competitors. Sometimes the old government-owned telephone services were privatized.

Removing the monopoly of the slow-moving, progress-resistant national telephone authorities coincided with the chip industry's capability to squeeze the electronics of a mobile phone onto half a dozen chips. The result was an explosion in the use of pocket telephones and of companies providing the networks on which they work.

Commercial competition meant that the prices of telephone calls – both mobile and conventional – tumbled all round the world in the mid-1990s, and the technological advance of microelectronics meant that equipment costs tumbled too. So the greatest effect of the chip on the telecommunications industry probably only started in the 1990s.

In other areas the pervasive, democratizing chip is bombing prices. TVs, video recorders, video cameras, printers, fax machines and mobile phones have been reduced from luxury items to commodities (Fig. 2.2), while radios, cameras, calculators, watches, and clocks have been reduced from commodity items to Christmas stocking fillers. This is simply explained. In 1970, the chip industry could sell you a chip with 1000 components on it. In 1994, it can sell you a chip with 16 million components on it for a comparable price.

The pace of change was not inevitable. If the technology had remained locked up in large corporations the pace would probably have been slower. But the history of the microelectronics industry is one of individuals taking the technology out of the large corporations and running with it themselves.

Because they were technologists first and businessmen second, these men created an industry whose economics often resembled the economics of the madhouse. In some ways they still do. The story of how they turned the chip into a global industry is in Chapter 3.

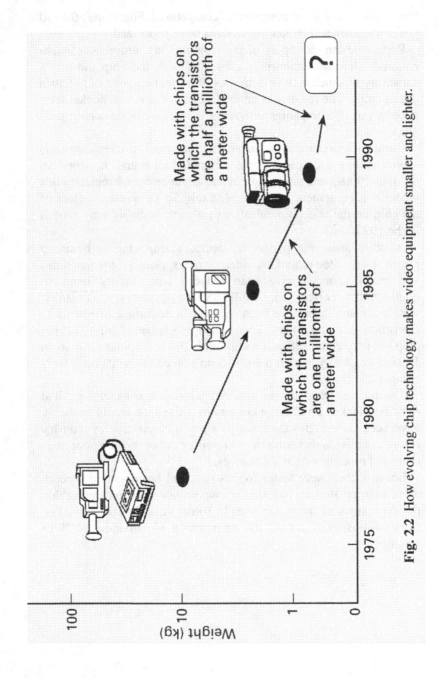

Fig. 2.2 How evolving chip technology makes video equipment smaller and lighter.

3

How the chip became an industry

'The business of America is business' – and the best endorsement of President Coolidge's maxim is the Americans' unrivaled capability at inventing new industries. That's especially true when it comes to taking science out of the laboratory and into the factory. So it was with the microelectronics industry.

If a seminal date is needed, then it must be 1955. That was the year William Shockley left Bell Labs and set up his own company to develop and manufacture transistors, having failed to persuade Raytheon that it was a good idea to pay him $1 million over three years as a consultant.

He named the company after himself – Shockley Semiconductor. Although other companies, both the tube manufacturers and the new companies specializing in transistors, were involved in transistor manufacturing before Shockley Semiconductor, no other company was to have such a profound effect on the way the microelectronics industry was to evolve.

Shockley had been brought up in Palo Alto, half an hour's drive south from San Francisco, and it was there that he decided to set up his new company with backing from Arnold Beckman, head of Beckman Instruments of New Jersey. Shockley recruited the brightest young researchers in the field, who were dazzled by his reputation. One of them, Robert Noyce, recalls 'It was like getting a phone call from God'.

Working for God had its ups and downs. On the upside, Shockley was not only an outstanding research director – quickly seeing the right direction for research and lucidly explaining it – he was also an entertaining boss, bubbling with enthusiasm himself and enthusing others. The day it was announced that he'd won the Nobel Prize for physics he took the whole company out for a champagne lunch at a Santa Clara eating place called Dinah's Shack.

He had a showman's way of illustrating his ideas, explaining the transistor's capability to boost a signal by comparing it to tying a bale of hay to a donkey's tail, lighting it and comparing the energy spent in striking the match to the energy spent by the donkey.

Shockley had two tricks for beginning his public speeches. One was to remark that his talk was on a hot topic whereupon he opened his notes and a puff of smoke came out. The other was to say he had only once before received such a flattering introduction – when the absence of the meeting's chairman had obliged him to introduce himself – then a bouquet of conjurer's flowers would spring up in his hand.

However, like Edison, Shockley had a downside. One of the researchers at Shockley Semiconductor, Gordon Moore, recalls: 'Shockley was an amazing man, good on problems, not so good with people. His physical intuition was unsurpassed – one colleague said "Shockley could see electrons" – but he had a peculiar idea of how people worked'.

Shockley saw plots everywhere and was paranoid about people stealing secrets. He fired staff on mere suspicion. On one occasion he became convinced that the explanation for a slow-moving project was that someone was sabotaging the work and subjected him to a lie detector test.

Eventually his eccentricities proved too difficult to live with and, in 1957, eight of his team walked out. 'The idea of setting up a company never occurred to us', said Moore, 'we thought we liked working together and hoped that a company might hire the entire group'.

The group eventually found a backer – the owner of a company making military cameras and instruments on Long Island – Sherman Fairchild. Although Shockley set up the first microelectronics company in what is now known as 'Silicon Valley', Fairchild was the company that both invented and commercialized the chip and established the highly individual culture of the chip industry.

Noyce, the leader of the defecting team, had set ideas on how to run a company. He had been brought up in the Protestant Congregationalist, mid-West, corn-belt town of Grinnell, Iowa where the values were the values of the Dissenting Protestant Congregationalist church and everyone subscribed to them.

Fundamental to the values of the Dissenters was a rejection of social hierarchy. Noyce was not religious, but the values of his upbringing went with him, and when he set up Fairchild he set it up as a community rather than a company and a community in which each individual internalized common values and goals.

The value system carried over to his private life. Years later, when he had more millions than he could count, Noyce was seen queuing at a Silicon Valley bank for a banker's check for $1.3 million to buy an aeroplane.

At Fairchild the outward and visible signs of Noyce's philosophy were: no offices – everyone had the same sort of cubicle; no company limos; no reserved parking spaces in the company car park; no dress code. Everyone had the right to speak up and the value given to an opinion bore no relation to the position of the person who made it.

Today, the presidents of new Silicon Valley companies assure you that their companies are communities rather than corporations, that everyone has stock and that everyone shares the values and goals of the founders and, most important, there are no reserved places in the car park. Pure Noyce!

The Noyce formula was more than just the expression of his upbringing, it was a practical way of getting the best out of the highly intelligent but egotistical band of scientists he gathered around him. The people were highly educated, mostly PhDs, and they were in a pioneering field. It was vital that every ounce of brainpower should be allowed to make its maximum contribution and that the demands of clashing egos were accommodated. An egalitarian community in which everyone had their say was an efficient way to achieve it.

Young graduate engineers were given unheard of responsibilities and left to pursue them without supervision. At Fairchild the purchasing policy was that engineers could buy anything they wanted – unless someone else objected!

Within a couple of years Fairchild had achieved what Edison used to call a 'Big Trick' – the invention, or rather co-invention, of the microchip itself. The immediate result was that Fairchild's East Coast parent, the Fairchild Camera and Instrument Corporation, exercised their right to buy out the shareholdings of the eight founders. Each of the eight got $250 000.

The Fairchild Camera and Instrument Corp. was not dumb: it realized that the chip opened the door to limitless possibilities for miniaturizing microelectronics. And it arrived at the very moment the Americans wanted limitless miniaturization. In 1961, three years after the birth of the chip, President John Kennedy committed the USA, which had been shocked by the Russian Sputnik launch of 1957, to putting a man on the Moon by the end of the 1960s.

Until then the US military had shown no interest in buying chips, which they saw as expensive and unproven. They preferred tried and trusted technology. Kennedy's commitment made them think again. Miniaturization, regardless of expense, was the new priority if the navigation and guidance systems required for a Moon rocket were to meet the size and weight specifications set by the designers of the Apollo rockets, which were to be the vehicles for the first manned Moonshots.

Then, in 1962, the engineers working on the guidance system of the Minuteman II intercontinental ballistic missile decided to adopt chips. Later that year the Apollo engineers made the same decision. In 1963 nearly all the chips made in the USA went to either the Minuteman II or Apollo programs.

Without the pressure to miniaturize irrespective of cost, chip technology might have been abandoned at this stage as too expensive to pursue. Which would have been a pity, because the technology's potential was about to be recognized in one of the most famous predictions ever made about microelectronics.

In 1964, Gordon Moore predicted that the number of elements on a chip would double every year. 'Moore's Law', as it is universally known to chip people, is still quoted as representing the essential nature of the pace of microelectronics evolution, although, since around 1970, the number of transistors in a chip has doubled almost every two years (Fig. 3.1).

Fairchild made two fundamental contributions: the first was to make routine the manufacturing of chips in large quantities and consequently at low cost. Just as important was the second: it set the pattern of technologists leaving one company to found their own companies.

Backed by unquenchable egos and venture capital from the financiers of San Francisco half an hour to the north up Route 101, a steady stream of technologists pursued the Silicon Valley start-up model. The great value of the model was that it gave technical autonomy in the new companies to engineers, and particularly to

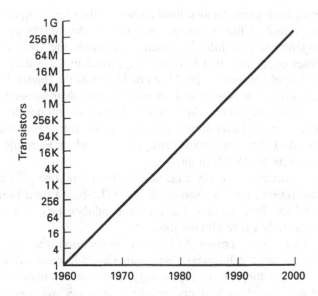

Fig. 3.1 Moore's Law: how many transistors you can put on one chip 1960–2000.

young engineers, because the great ideas that advanced the industry tended to come from engineers in their twenties and thirties.

In large companies such people would have had little influence over strategy; big company controls meant that their ideas would have been ignored until they had become obvious – and by then it would have been too late to use them. As long as the industry's expertise was owned by people rather than by companies, people could leave companies and prosper in the industry.

The model of engineers splitting away from companies to start new ones persists to this day in Silicon Valley and it makes the Valley the world's intellectual HQ for the semiconductor industry. Most of the major innovations in the technology have come from Valley companies, and when companies and countries want to get into the semiconductor industry today, they commonly buy into or buy up Silicon Valley companies.

Shockley Semiconductor, for instance, had been bought by Clevite in 1959, two years after the eight defected. In 1965 Shockley Semiconductor was sold on by Clevite to ITT.

For Fairchild, the first defection of engineers to start a new company came in 1959, only two years after starting up. In 1961 there were two more defections. In 1963 came another. Then came defections from the defectors, until a chart of the chip companies in the Valley looked like a complicated family tree springing from one progenitor – Fairchild. So, naturally enough, the 150+ micro-electronics companies that have been spawned in the Valley since the founding of Fairchild in 1957 became known as the 'Fairchildren'.

Eventually, even Noyce and Moore became disillusioned with their Long Island parent, which was siphoning off money from its chip subsidiary and investing it in areas outside the semiconductor industry. And the final straw came in 1968 when Fairchild was looking for its third CEO in one year.

'The chairman was really a bit of a buffoon – no one paid much attention to him', recalls Gordon Moore. 'The board fired him and then hired and fired another. The logical candidate was Bob Noyce but they wouldn't give him the job.'

The East Coast bosses had never understood the way the Californians ran the business. 'Sherman Fairchild required all his people to go through a psychology test', remembers Moore. 'Noyce and I read each others' results – they said we were good technically but would never be managers.'

To Noyce and Moore, the East Coast people had begun to represent traditional, US corporate life – self-indulgent executives having long lunches. Corporate jowls and paunches were not the fashion among the 30-year-old CEOs of Silicon Valley.

Noyce and Moore, with Moore's deputy head of R&D Andy Grove, quit with the intention of setting up a company, which they called Intel. The aim was to pioneer the idea of using chips to provide the memory in computers. At that time magnetic cores were used for storage and Noyce reckoned chips could reduce the cost of memory storage 'by a factor of one hundred'.

The technology that they intended to use to achieve that goal was called MOS (Metal Oxide Semiconductor) – then a laboratory concept, nowadays the workhorse production process of the world-wide microelectronics industry. 'MOS had the right degree of difficulty', says Moore. 'It was seven years before we had any competition – that gave us a chance to expand in a vacuum.'

Three years after being founded, Intel had invented all the basic types of memory chip that are used today – dynamic and static random access memories (DRAMs and SRAMs) and electrically programmable read only memories (EPROMs). In the mid-1990s these products account for about a fifth of the chip industry's total sales – a proportion that is steadily increasing.

However, in 1971, Intel brought out another invention with greater potential the microprocessor. The chip could be programmed to operate as the basic control for almost any electronic product, but today it is principally used as the 'brains' of a computer.

Because of its wide use, the microprocessor could be made in high volume and consequently cheaply. Its inventor Ted Hoff claimed that it 'democratized the computer', bringing computers into the realm of a High Street shopping item rather than a multi-million pound corporate tool operated by scientists in white coats.

MOS memories and microprocessors were the foundation for Intel's rapid success. In 1971, three years after being founded, revenues topped $9 million, and by 1976, they were over $200 million.

Having a second chance to start a company from scratch allowed Noyce, Moore and Grove to do things their way. All the engineers and most of the office staff got stock options, which was something Noyce had wanted at Fairchild but which the East Coast parent had refused to allow.

Stock options were not just a good way of motivating staff by linking their own finances to the company's fortunes; they were also a good way of fixing in employees' minds that research breakthroughs – the lifeblood of Silicon Valley companies – resulted in a leap in the share price. That way the employees were keener on pursuing breakthroughs than they were on making profits from existing products – a key stimulus in focusing the whole company on pushing forward into newer, higher return areas.

Naturally there were no offices, no hierarchy and no reserved parking places, but at Intel they went even further than at Fairchild, with open plan offices where everyone could see everyone. Noyce himself worked at a beat-up secondhand metal desk and, as the company grew and new desks were bought for new staff, even the typists had bigger and better desks than his.

As the company grew, sessions explaining Noyce's culture to the newcomers became regular events. A key to the culture was that everyone was supposed to speak up if they had an idea. No matter if it contradicted Noyce or Moore or Grove, the rule was to speak up and challenge them. This was heady stuff for the bright young engineering graduates recruited by Intel, and vital in making them self-reliant, self-confident and capable of taking initiatives on their own responsibility.

By the company's fifteenth birthday it had reached a billion dollars in sales and in 1993 it was the world's largest chip company, with around $8 billion worth of sales. In the mid-1990s Intel's sales of microprocessors were topping 40 million microprocessors a year at prices ranging from $1000 to a few dollars a piece, and its microprocessors powered over 80% of the world's personal computers. Gordon Moore, who, according to Sherman Fairchild's psychology test, 'would never make a manager', was still Chairman.

Besides Intel, two of the other companies founded by technologists leaving Fairchild became billion dollar companies – Advanced Micro Devices and National Semiconductor. The great value of the Silicon Valley system of companies spawning new companies was that it made constant innovation the basic ingredient for success.

Only by getting a technical lead on the competition could a company prosper. Since established companies are usually slow to back new ideas, the Fairchildren had the opportunity to outrun the competition by pushing the technology. For America it was a win–win situation, as innovation was driven forward at a dizzying pace, but for the older established companies it spelt disaster.

In 1955 the top ten makers of transistors in America (and also in the world) were Hughes, Transitron, Philco, Sylvania, General Electric of the US, Clevite, Westinghouse, Motorola, RCA and Texas Instruments. Twenty years later, in 1975, all except the last three had been replaced by Fairchild, National, Intel, Signetics, General Instrument, American Microsystems Inc. (AMI) and Rockwell (of which five were Fairchildren founded in the 1960s).

Today the start-up rate of new companies has not slowed down. Below the top companies are a raft of smaller companies in all sizes from half a billion dollars to a few thousand. Their combined output, worth several billion dollars annually, makes up a sizable proportion of America's total output of chips.

Moreover, the small companies are responsible for some of the major innovations, such as the 1990s trend towards logic chips that are customized by the user rather than by the manufacturer. Programmable logic chips represented a major change in the industry – the biggest since Intel invented MOS memories and microprocessors – and the concept was pioneered by start-up companies.

The mechanics of the Valley are such that the people with good ideas get a chance to run with them. All the basic ingredients for founding a chip company – financing, knowledgeable lawyers, sympathetic estate agents, helpful suppliers and a host of support services – are all on the doorstep of the newly hatched microelectronics company. As one of the early microprocessor pioneers, Federico Faggin (designer of the 8008, 8080 and Z80) puts it: 'The Olympics of entrepreneurism are held here'.

Having invented the chip industry, the Americans dominated it throughout the 1960s and 1970s. Europe and Japan were two to three years behind the Americans and no one else was even in the race. The top ten American chip companies were the world's top ten chip companies.

In the 1990s things are different. Towards the end of the 1980s the Japanese started to out-manufacture the USA in chips, and there have been more Japanese companies than US companies in the top ten for the early part of the 1990s. Behind that switch lies a remarkable effort by the large Japanese electronics companies.

In the 1950s, the Government of Japan took legislative measures to support or develop microelectronics in its country. It did this in three basic ways: by putting steep tariffs and restrictive quotas on the import of advanced chips; by requiring American companies wanting to sell chips into the Japanese market to license their technology to local companies; and by requiring Japanese companies which obtained such licenses to sub-license the technology to other local companies to spread technological understanding as widely as possible.

However, the Japanese government's attempts to acquire a national microelectronics capability were not always smart. 'MITI has not been the great benefactor of the Japanese electronics industry that some critics seem to think it has', said Akio Morita, founder of Sony, in his book *Made in Japan*.

Morita tells how MITI (Japan's Ministry of International Trade and Industry) had delayed his efforts to license transistor production know-how in 1952. That year the transistor's inventors, AT&T's Bell Labs, held a symposium to reveal how transistors were made to anyone prepared to stump up $25 000.

Morita and Sony co-founder Masaru Ibuka had recognized the potential of the transistor and were particularly keen to learn how to make it. However, MITI used the exchange control regulations then in force in Japan to try to persuade Sony to give up sending the $25 000 to AT&T.

'The bureaucrats at MITI could not see the use for such a device and were not eager to grant permission', said Morita. It took Ibuka six months to get MITI to change its mind. Eventually MITI relented and Ibuka went off to the USA to acquire the keys to what is, arguably, the most important industrial technology of the second half of the 20th century.

After substantially re-engineering the Bell Labs transistor, in the course of which a Sony engineer Leo Esaki discovered a phenomenon called 'electron tunneling' that won him a Nobel Prize 20 years later, a transistor was produced that was suitable for a portable radio.

However, the big electronics companies were conservative about investing in the microelectronics industry. As late as 1968, total Japanese chip production was worth under $25 million with 10% of those revenues going on royalties to American companies. Meanwhile, the combined R&D spend on semiconductors of Japan's three largest computer makers, Hitachi, Fujitsu and NEC, was less than the R&D spend of the leading American microelectronics company, Texas Instruments.

One move the Japanese government did make at the end of the 1960s was to liberalize trade and foreign investment in the chip industry, allowing direct investment by foreign companies in the Japanese chip industry.

For many years, attempts by Texas Instruments, then the world's No. 1. chip-maker, to get permission to build a chip factory in Japan had been resisted. In 1967, Sharp of Japan wanted to export portable calculators to America and Texas strongly opposed the move on the ground that Sharp's calculators contained chips that infringed the patent held by Texas' Jack Kilby on his original chip. The pressure paid off and, in 1968, Texas built its factory in Japan.

By relaxing these restrictions and allowing American companies to operate in Japan, the government achieved three things: it helped to defuse trade friction with America, which was a problem even then, it helped to build up a local core of trained workers in the chip industry, and it prepared the scene for the day when the Japanese chip industry would be able to compete with the American chip industry on a basis of equivalent technological capability.

By 1970, MITI had come to the conclusion that if it was to achieve its ambition of success in the computer industry it had to do more to achieve a leading capability in microelectronics.

The thrust to achieve that equivalence with the USA was spearheaded by a famous, but controversial, research project which came to be known as the VLSI (very large scale integration) programme.

It was so named after the body spearheading the effort, called the VLSI Technology Research Association, which drew together the top technologists from all the major companies. Some of the top companies protested that the project was unnecessary and that they could develop the technology on their own. One reason for the protest was that they did not want to share their technology secrets with other companies which they saw as rivals.

As one senior Japanese chip industry executive put it at an international conference in 1991: 'The VLSI program did not accomplish anything by collaboration that we could not have achieved separately by ourselves. Except one thing – it educated company presidents in the importance of microelectronics'.

However, MITI usually gets what it wants in Japan and the VLSI program got started in 1976. The main companies got together in two groupings to pursue the research. One group was Hitachi, Fujitsu and Mitsubishi, the other group was Toshiba and NEC. It cost ¥70 billion, ¥30 billion from MITI and ¥40 billion from the companies involved.

Quite how much of the subsequent success of Japan's chip industry can be attributed to the VLSI program cannot be estimated, but, in 1983, NEC and Hitachi began mass manufacturing the most advanced version of the most widely used memory chip – a DRAM capable of storing 256 000 binary digits – a full year before any American company could produce it.

Two years later the financial muscle of Japan was shown when a recession in the chip business – which is subject to erratic fluctuations in supply and demand – caused, according to American industry estimates, a worldwide industry loss of $6 billion. But whereas all the US companies dropped selling DRAMs – the most expensive product to produce and the most widely sold – all the Japanese companies stayed in.

Meanwhile, between 1978 and 1988 six American producers dropped out of the ranks of the world top ten – RCA, Fairchild, General Instrument, AMI, Rockwell and National Semiconductor. They were replaced by six Japanese companies: NEC, Toshiba, Hitachi, Fujitsu, Mitsubishi and Matsushita.

Nowadays Japan and America are level-pegging in chip manufacturing, although the Americans have seen a resurgence in their market share as they have moved to higher value-added products, leaving the commodity end to the Japanese. And, in their turn, the Japanese are looking to higher value products as the South Koreans take an increasing share of the commodity end of the market.

One thing that helped the Japanese was that the companies involved were large and vertically integrated and were focused on the consumer industry, which required the development of low-cost and low-power technologies. This was a very different priority from the American focus on space and military applications.

Another positive factor for Japan was that her major electronics companies were substantial equipment companies able to subsidize unprofitable ventures with profits earned elsewhere and capable of sustaining a long-term strategy.

Furthermore, what allowed Japanese companies to make the large capital investments required for success in microelectronics was that interest rates in Japan were low in the 1970s and cheap capital could be borrowed readily because the Japanese government implicitly guaranteed loans made to strategic industries, of which microelectronics was one. Accordingly, many microelectronics companies in Japan have debt-to-capital ratios in the 60–70% range, compared with the sub-20% commonly found in America.

This is something that is changing. American executives are expecting the cost of capital in Japan to be higher than that in America in the mid-1990s, which will further level up the playing field.

In the early 1990s, Intel out-invested every other company in the industry, and in 1992 overtook NEC of Japan to become the world's largest manufacturer of chips. So the USA/Japan battle for supremacy in the chip industry is far from over.

Europe still hasn't got the hang of manufacturing microelectronics successfully. Even after four decades of governmental support in the forms of financial assistance and protective tariffs, Europe produces less than 10% of the world's output of chips; a poor achievement when it is remembered that Europe has been involved in the chip business ever since it started.

The lack of success is probably because European electronics is dominated by big independent-minded companies which can't be influenced by governments as effectively as in Japan and because Europe is without a tradition of technological entrepreneurism, as in Silicon Valley.

In Germany there are two significant players, Siemens and AEG. Siemens spent $200 million in the 1970s to boost its microelectronics capability with $40 million coming from the German government. In the 1980s it met the increasing costs of microelectronics research and development by combining with other companies in a series of cooperations to make advanced memory chips.

First it combined with Philips in a billion dollar program called 'Megaproject'. Then it went in with Philips and SGS-Thomson in a $4 billion project called 'JESSI' (Joint European Submicron Silicon Initiative), then with IBM. The partnerships have been subsidized variously by the German, Dutch, French and Italian governments and the European Commission.

In the early 1990s Siemens continued the process of collaboration by embarking on a billion dollar joint research and development effort on 256 megabit memory chips with IBM and Toshiba and looked around the world for manufacturing partners. By then, most of the world's top chip companies had decided that collaboration was the way to go.

AEG has always avoided the mass market and operates in a series of niche markets, which are more profitable than the mass market but which do not require the same commitment to staying at the leading edge of the technology. In 1990, AEG amalgamated all its niche chip activities into one company called Temic – an operation with enough critical mass to attack mainstream markets.

Philips of Holland has been the most consistently successful European microelectronics company over the years. It has chip manufacturing operations worldwide and technology exchange deals with Intel. However, its partnership with Siemens in memory chips did not turn into a market presence in such chips and without the need to have top-class chip capability to support its consumer electronics manufacturing Philips would probably invest a great deal less in microelectronics.

After Philips, the second biggest European microelectronics company in the mid-1990s is the Franco-Italian company SGS-Thomson. From its foundation to the mid-1990s the company received generous support from the French and Italian governments and its specialities – such as EPROM chips – received Euro-funding for R&D and a protected trading environment through EU (European Union) political moves. Along with Siemens and Philips, SGS-Thomson is one of only three European companies that have a leading-edge microelectronics capability.

The UK has agonized over the chip industry from its earliest beginnings. Having failed to back the ideas of Geoffrey Dummer of the Royal Signals and Radar Establishment (RSRE), who had produced the world's first model of a chip at the International Symposium on Components in 1957 – a year before Kilby – the government belatedly backed the UK's fledgling chip industry with R&D money.

In the 1960s the UK government funneled money into Marconi-Elliott Microelectronics, Plessey and Ferranti. Indeed, in 1967 it gave Elliott Automation a contract to instal an MOS process (a year before Intel was founded expressly to develop MOS). Government money for chip research was $8 million in 1968 – about a third as much as America was spending.

However, a government-sponsored reorganization of the UK electronics industry in the late 1960s saw mergers between GEC, AEI and English Electric, which had severe repercussions on the UK microelectronics industry. In 1968, GEC sold off most of its shares in the UK's second largest microelectronics company, Associated Semiconductor Manufacturers (ASM) – a joint venture with Philips.

Then in 1971, during one of the periodic industry downturns, GEC closed the chip factories of Marconi-Elliott Microelectronics and Elliott Automation, which had been making standard logic chips under license from Fairchild. That put the UK out of the market for standard chips.

Seven years later GEC tried to get back into the standard micro-electronics business via a joint venture with Fairchild. Fairchild was by then long past its glory years, Noyce and Moore having left a decade previously. The venture was abandoned after Fairchild was taken over by the French oil-field services company Schlumberger.

Ironically, although GEC was preferred by the government rather than Plessey as the lead partner to amalgamate the ailing AEI and English Electric – so making GEC the only company with the critical mass to sustain a commitment to high volume microelectronics manufacturing – it was the smaller Plessey which was to show greater commitment to microelectronics over the succeeding years.

The UK government had two last throws at trying to get a micro-electronics industry established. In 1978 it backed a new chip company called Inmos with an eventual $176 million. Two of the three co-founders were American technologists.

Inmos had inspired ideas but poor execution. Its most brilliant idea was an ahead-of-its-time microprocessor which it called the 'transputer'. However, the transputer was difficult to use and not oriented to the mass computer market, and has remained a niche market chip. Eventually Inmos and its transputer were sold off to Thorn, which passed it on to SGS-Thomson.

The other 'last throw' by the UK government to get a strong microelectronics industry going was a 1978 $200 million five-year collaborative research program called the 'Alvey' project. If its aim was a thriving UK chip sector, it failed. A decade later the chip-making operations of GEC, Plessey and Ferranti had all been folded into one medium-sized chip company.

If the British failed at microelectronics manufacturing, they had the consolation that the French did little better. In the 1970s a series of moves aimed to keep a national chip-making capability: Alcatel, the telecommunications giant, set up a medium-sized chip subsidiary, Mietec, mainly for supplying its own needs; Matra, the defense company, went in with the US company Harris on a joint venture which was eventually acquired by AEG; National Semiconductor was engaged in a joint venture called 'Eurotechnique' with the industrial combine St Gobain only for it later to be sold off to Thomson; Thomson was sustained by the government more or less generously until it was merged in with SGS of Italy and is still in receipt of the French taxpayers' franc; and most dramatic of all the French failures in microelectronics was the purchase of Fairchild by the French oil-field services company, Schlumberger, in 1978 for $425 million. Schlumberger pumped in a further $1.3 billion over the next eight years before selling it to National Semiconductor for $122 million.

Eastern Europe emerged in the mid-1990s from its position as a supplier of standard memories and microprocessors to other East European companies and, mainly, to the military. It had to undergo a painful transition period as it adjusted to operating in world semiconductor markets, but four Russian companies – Mikron, Angstrem, Electronica and Integral – showed themselves up to the task, initially exporting watch, calculator and games chips and ASICs (applications specific integrated circuits) while refining their expertise in higher value-added areas.

Many people have asked why Europe has so signally failed in the microelectronics industry despite the lavish government support. Maybe it takes an outsider to see the answer. 'It's not that

European research and development isn't good', reckons Gordon Moore. 'The lab work in Europe was always outstanding, but it's just not translated into manufacturing.'

Apart from America, Japan and Europe there are only two other places on Earth where there is a significant capability in advanced microelectronics manufacturing – South Korea and Taiwan. Each has gone about acquiring its chip-making ability in a totally different fashion.

It may not be too fanciful to say that microelectronics research, development and manufacturing is such a difficult thing that the national character of different regions shows through in the way they approach it.

With the Americans it is individualistic – strong characters doing their own thing in their own way and in their own companies; with the Japanese the important thing is the art of working together successfully in groups; with the Europeans it is getting so involved in trying to understand the science that nothing much gets made.

If the national character theory holds good, then the South Koreans and the Taiwanese fit the theory. Their approaches to getting chip technology are totally different. The South Koreans have bulldozed their way to a leading edge capability with a massive input of money and a relentless exploitation of every avenue to acquire the capability.

In the 1980s, when South Korea decided to become a major player in the microelectronics industry, they built massive factories in South Korea, they set up companies in America to recruit clever technologists to pick their brains, they bought Silicon Valley companies to acquire their technologies, they licensed technology and products from the Americans and Japanese and they manufactured products for other companies.

Between 1983 and 1989 the South Korean output of chips went up from $500 000 worth a year to $1.5 billion – an unprecedented surge in output. Now three South Korean companies, Samsung, Goldstar and Hyundai, have sizable and advanced microelectronics capabilities.

The most successful of the three, in the mid-1990s, was Samsung, which astonished the microelectronics industry by building three factories simultaneously to make three generations of chips. The normal practice in the industry is to use the profits from one generation to fund the construction of factories to build the next.

There is no way of knowing the extent to which the South Korean government funded the effort or how much came from the World Bank in low-interest loans. However, the government provided support on two fronts. In 1979 it set up the South Korean Institute for Electronics Technology (KIET) with $60 million in backing to develop technology and transfer it to the industrial companies, and in 1982 the South Korean government announced the 'Semiconductor Industry Promotion Plan' aimed first at import substitution, then at exports.

Their subsequent success in exporting has been met by the ultimate acknowledgement that they are competing successfully – anti-dumping actions in America and Europe initiated by the competition.

South Korea's success in the microelectronics industry may have surprised many people, but it has not surprised the Japanese, their Asian neighbors, who know the South Korean character. The people working in the South Korean companies at achieving the capability have seen it almost as a crusade, certainly as part of a vital and patriotic national duty. Dedicated effort has been a vital factor in South Korea's success.

In Taiwan the technology was acquired by a very different route. Back in the mid-1970s the island's government decided it needed microelectronics technology to support its ambitions in the electronics industry.

In 1976 the Taiwanese government research body, the Industrial Technology Research Institute (ITRI) bought from RCA the technology to set up a microelectronics laboratory in Hsinchu City, which is an hour's drive south of Taipei.

The laboratory is still there. The only difference is that whereas in the 1970s it was making transistors seven millionths of a meter across, in the 1990s it is now making transistors as small as a quarter of millionth of a meter across, which is as advanced as any commercial process in the world today.

At the same time as ITRI's lab refined the RCA process year by year, the electronics arm of ITRI, called ERSO (Electronics Research and Services Organization), set itself to building factories every five years or so which were capable of reproducing the laboratory process in a manufacturing environment. When the factories were built and a new level of process technology had been developed, the team of technologists that had developed the process in ITRI's lab would move over to ERSO's factory to get the process set up.

On three occasions the ERSO factory has been handed over to a new company to be its first manufacturing facility and, each time, many of the ERSO technologists moved over to work for that company. The first time it happened was with United Microelectronics Corporation (UMC), Taiwan's largest chip-maker; the second time was with the Taiwan Semiconductor Manufacturing Company (TSMC), and the third time with Winbond Electronics Corporation.

The ITRI formula for hatching out microelectronics companies appears to have been very successful. That is probably because it addresses one of the most difficult problems in the microelectronics industry – how to transfer microelectronics technology from the laboratory to the factory.

The ITRI formula was backed up by a policy of sending its brightest engineering students to American universities, many of whom subsequently work for US chip companies before returning to Taiwan to join local companies.

During the late 1980's stock market boom in Taiwan, when raising money was easy, many of these US-educated Taiwanese engineers dashed back to the island to share in the boom. 'They only had to get off the plane from the US and make one phone call to have the postman dump $20 million on them', recalls Klaus Wiemer, then President of TSMC.

The net result was that, by the mid-1990s, Taiwan had six microelectronics companies in the $100 million+ league, which is more than any country in the world apart from America and Japan. Moreover two of them, UMC and TSMC, started building billion dollar chip factories in 1994 – factories on a scale only affordable by the world's largest microelectronics companies – signaling their determination to become major league players during the 1990s.

The Taiwan top six, in addition to UMC and TSMC, are Winbond, Hualon Microelectronics Corporation, Macronix and Mosel-Vitelic. In addition to these there are a host of small design houses in Taiwan, capitalizing on the Chinese flair for chip design.

The story of how the chip was born by the Americans, adopted by the Japanese, intellectualized over by the Europeans, bulldozed into play by the South Koreans and patiently nurtured by the Taiwanese parallels the post-war rise and fall of nations.

The entry port to microelectronics is money; the necessary ingredients are a society with a high level of scientific and technical education and understanding married to a dynamic to improve itself. Perhaps the interesting thing is that, in nearly four decades since the chip was born, so few countries in the world have mastered the black art of manufacturing it.

4

Where are we now?

Notwithstanding the enormous impact of the chip industry in its first 35 years, that period could be regarded as the industry's adolescence. For the public at large the full flowering of the chip's enormous potential did not start to become apparent until the mid-1990s.

In those first 35 years, the chip went from having a couple of transistors to 64 million. In doing so, it bombed the price of a whole range of electronic products – some to the point of becoming giveaways.

None the less, they were 35 years of preparation in which people became aware of the power and potential of the chip. Not until the mid-1990s did the chip really achieve the necessary cost/ performance capability to start fulfilling its enormous potential.

And it was not until the mid-1990s that the developed countries started to put the necessary political and physical infrastructures in place for the chip to really exploit its great power.

In the mid-1990s we are embarking on a major change in the way the developed world operates, because three commodities that we are used to paying a lot of money for are going to become virtually free. These are information storage on chips, information processing by chips and the transmission of information by chips.

The reason why the 1990s will see an explosion in the capabilities of electronics equipment – based on cheap chip storage, processing and transmission – is that, while the doubling up capability of microelectronics technology roughly every two years for about the same cost is continuing, we are now doubling up in tens of millions of transistors rather than in the thousands of the past. And, starting with the quarter-billion transistor chips due on the market in 1997, we will be doubling up in billions.

For the first half of the 1990s, state-of-the-art commercial chips being made in large volume contain 16 million transistors – enough to store some 700 single-spaced typed A4 pages or a copy of *Gone with the Wind*.

In 1994 these 16 megabit memory chips cost from around $50, and industry analysts forecast they will probably decline in cost by 30% a year over the first few years of their probable ten-year lifetime. Meanwhile, chips from the previous generation of 4 megabit chips sell for four times less (around $12) and chips from the generation before that – the 1987 vintage 1 megabit chips – sell for a quarter of the price of a 4 megabit chip (or $3).

Although supply/demand imbalances can artificially depress/increase prices in the short term, the long-term trend in the industry has been for a 30% decrease in price per year.

Bearing in mind that a 1 megabit chip is no shabby piece of technology – it can still store 45 pages of A4 single-spaced typing – the statement that chip memory is becoming virtually free is permissible.

So at the bottom end of the electronic products business, ever-cheaper chip capabilities are bombing prices remorselessly. On the other hand, at the top end, the capability of chip technology to do more and more continues to grow exponentially and, by 1996, commercial chips will have 64 million transistors.

Since each chip eventually comes to cost just a few dollars after only three or four years from its first introduction, that means huge decreases in cost-per-function are on the way.

So the size and cost reduction process from now on will deliver major leaps ahead in technological capability to the manufacturers of electronic products. On the one hand that means that every ten years we will see the price of the latest and best electronics products at the start of the decade drop to the point where they are commodities by the end of the decade. On the other hand, it means that the latest electronics products of the next decade are going to have quite unexpected capabilities.

For instance, ease of use of electronic products is one big next step for chip technology to crack. Few of us use the full capabilities of our computers or even know the function of every button on our TV remote controls. As for programming video recorders – many of us don't even try.

It would be a big benefit to us all if the machines could themselves show us how to use them and make it simple for us to do so.

Quite soon there will be screens on products like video

recorders, copiers, faxes and printers. These screens will tell you what the machine can do, how to operate it, where you've gone wrong and how to put it right.

Later on this decade the machines will be able to tell us what to do in human language. We'll ask them to do things and they will be talking back telling us how to use them or what's wrong with them.

To add those capabilities to the machine means, in effect, putting more of the intelligence into the machine while expecting less intelligence from the person using it. And the way to put more intelligence into the machine is to put more memory chips and microprocessors into it.

Since people always require more added value in the product for any increase in cost, then it is only the declining cost and increasing performance of the chips that makes it possible for the video or printer manufacturers to add this extra intelligence without it costing more than the consumer is prepared to pay.

So putting more and more intelligence into the machine is a key need for the 1990s and one which cheaper and cheaper chip capabilities make possible. The more transistors you can affordably use in a product, the more intelligence you can build into a product. And the more intelligence you can build into the product, the easier it can be to use and the greater the number of people who will want to buy it.

Which gives us the traditional microelectronics industry virtuous circle: as the cost per transistor drops every year, and the size of the transistor decreases every year, so it becomes cheap enough and small enough to provide more transistors in the products to make them appeal to more people.

Speed is another increasing capability for the computer. The more transistors there are on a chip, the closer they are together and the less distance the electrons have to travel between them. The less distance they have to travel, the faster they do the job, and the faster they do the job, the more work they can do in a given time.

The chip that decides how fast a computer works is the microprocessor chip, and microprocessors have been doubling in speed every two years. From the 50 million instructions per second chips of the early 1990s, the end of the decade should see 1000 million instructions per second chips.

That represents the kind of power needed for performing very complicated tasks like the instantaneous translation of one spoken language into another. Once that becomes possible to do on one chip, the inexpensive portable translation machine – a real dream machine for the world – will become possible (Fig. 4.1).

As well as making new things possible, the chip also makes existing things cheaper. One way in which the electronics industry is about to explode is that the computer is becoming affordable to everyone.

About 50 million computers were being sold every year in the mid-1990s – peanuts compared with the 5.6 billion population of the planet. As a product which should halve in cost every two years the personal computer has hardly made a dent in its potential marketplace.

Seven-year-old chip memory used in computers is, as we have seen, almost a free commodity. Similarly, simple microprocessors of mid-1970s vintage can perform many of the tasks which most people want a computer to do – e.g. word processing, spreadsheets and games – and these old microprocessors now cost only two or three dollars. So as well as delivering almost-free memory, the chip industry is delivering almost-free processing power.

Already the personal computer is outselling the car in unit volumes on a worldwide basis, and it is closing the gap with the worldwide unit volumes of TVs. As computers get to be able to handle more and more types of information – from phone calls to faxes to videos – there are going to be few citizens of developed and developing countries who won't want to own one.

And, as prices decline, more and more people are going to be able to afford them. There is very considerable potential for cost reduction of PCs. The keyboard cost may have hit rock bottom, but the other two main elements – the screen and the chips – can reduce dramatically.

In the mid-1990s, black and white screens were still costing around $100 for 30 lines deep and $150 for 60 lines – but they have the potential of reducing substantially because the materials involved are extremely inexpensive. Once production volumes are high enough, prices will drop like a stone.

The calculator and electronic watch markets dropped to give-away status when component prices fell. Calculators and watches use much smaller screens than computers, but they are made from the same materials using almost the same technology. For computer screens the same cost fall is likely to happen. It will trigger a big fall in the price of computers.

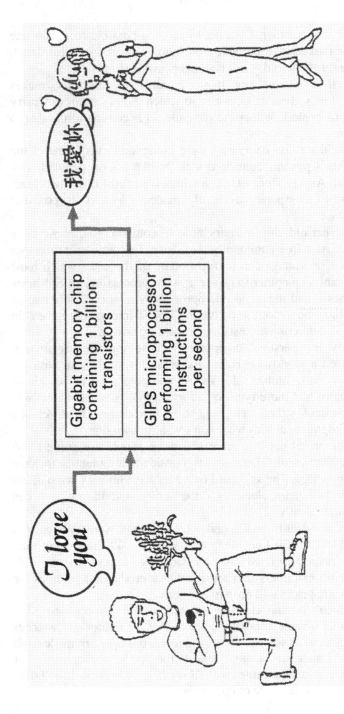

Fig. 4.1 Example of an electronics system in the year 2000: the portable translator.

For the chips, the price decreases every year. Moreover, chips should, by the beginning of the next century, start to take over from magnetic disks, becoming the only means of memory storage inside a computer. If that happens it will remove the bottom limit from the potential price decline of computer memory because the revolving mechanisms and motors required by disks are never likely to cost much less than $100, while the cost per bit of a chip has no theoretical bottom limit.

With the cost of memory chips, processing chips and screens dropping through the floor, personal computers are set to spread like weeds, to the great benefit of consumers around the world. But that's not the end of it: as well as giving us almost-free memory and almost-free processing, the explosion in chip capability is also on the verge of bombing the price of communications.

Just as potentially ubiquitous as the portable computer is the portable telephone. These are cheaper than computers and should soon be very much cheaper. Two-chip pocket telephones should be on the market in the mid-1990s which will have a major effect on reducing cost. Single chip phones – when the price really starts to bomb – should be around by the end of the 1990s.

However, the problem in spreading pocket telephones across the globe is not going to be the cost of the phone, it's going to be in setting up the networks. Although it is possible to cover densely populated areas like Japan or the UK with transmitters, each one serving an area sharing borders with other areas – like cells in a honeycomb (the so-called cellular telephones) – this is not cost-effective in the wide open spaces of countries such as the USA, Russia, India and China, where the population is scattered over a larger area.

To provide these very large countries with networks for portable phones requires an alternative to cellular telephones. There are two main alternatives: one is a network of low-orbit satellites to provide a comprehensive global network – a scheme currently under way through two international consortia, Iridium and Inmarsat. The other alternative to cellular is to use a combination of radio and telephone wires – using radio to and from the nearest telephone exchanges and telephone wires between the exchanges.

The problem in Japan and America has been getting the governments to allocate frequencies and set technical standards for the networks. Europe has, unusually, led the way in being the first major region to set frequencies and standards. But though Europe has set the standards for telephone calls for speech, it has yet to do so for radioing data.

When governments do set the standards and frequencies for wireless data communications there will be some explosive possibilities for worldwide communications. That's because chip technology now has made it inexpensive to digitize, store and transmit any form of data, from words to video.

For instance, the likely European standard for transmitting digital wireless data is that it will kick off at the high end of the transmission rate of the mid-1990s for data sent down telephone wires using equipment you can buy in the shops – i.e. around 24 000 bits per second. However, it is possible to see this being upped to a million bits per second during the course of the 1990s.

Since it requires 16 million bits to store 700 typed A4 pages or a copy of *Gone with the Wind*, it is going to be possible to send (at a million bits per second) an entire digitized version of *Gone with the Wind* from one portable computer to another (or to a fax machine or a printer) in 16 seconds.

In fact it will take less than 16 seconds because it is possible to précis digitized information so that it uses fewer bits of information storage space. These 'compression' techniques allow you to compress digitized written data by up to three times.

The technique of compression is theoretically pretty simple. It involves looking for repeated patterns in the 0s and 1s of digital language and replacing those repeated patterns with a symbol. Since compression can reduce the number of storage bits needed by up to three times, it is possible to store a compressed version of *Gone with the Wind* in just over five million bits. Transmitting that at a million bits per second will take only five seconds.

At the fastest rate available over telephone wires in the mid-1990s – 24 000 bits per second – sending five million bits will take about three minutes, which can involve, over long distances, a not insignificant cost.

But new techniques are increasing the transmission speed dramatically and, during the 1990s, one million bits per second and even several million bits per second transmission rates are likely to be achieved. At the same time deregulation and competitive pressures in the telecommunications industry are slashing the costs of telephone calls. So it can fairly be said that the cost of transmitting digitized data is heading towards becoming free.

Over shorter distances, for instance the telephone wires inside an office, transmission of data at ten million bits per second is possible in the mid-1990s. That would allow the transmission of a digitized, compressed, *Gone with the Wind* in 0.6 seconds. The rate of transmission in these circumstances (called in the industry a local area network or LAN) is set to increase at least ten times over the 1990s.

The transmission of other forms of information is not yet free but is getting that way. For instance, video pictures can also be digitized and compressed and sent down telephone wires or over wireless networks.

A single frame in a video needs two million bits of storage space on a memory chip. In a video film there are 30 frames for every second of film. So every second of film time requires 60 million bits of storage, which means that a one and a half hour video film can be stored in 324 000 million bits.

In the mid-1990s, using standard equipment bought in your local electronics store, the fastest you could transmit that one and a half hour video would be 24 000 bits per second. At that transmission rate it would take 150 days to transmit the one and a half hour video.

But the ability to compress video is very much higher than the ability to compress written data. Video can be compressed 100 times without losing perceptible quality. Therefore a one and a half hour video can be stored in 3240 million bits. Sending this down a telephone wire at 24 000 bits per second would take 1.5 days.

However, when the transmission rate is increased to a million bits per second, then you'll be able to send a compressed digitized hour and a half of video down a phone line in 50 minutes. Furthermore, it is not unlikely that 6 million bit per second transmission will be available in the 1990s, so reducing that time to a little over eight minutes.

The means for much faster rates of data transfer is, in the mid-1990s, being put in place. Optical cables, which use flashes of light to carry signals instead of the electrons which carry signals down ordinary wires, are being installed in every developed country.

Current optical cables deliver data at 2400 million bits per second or 2.4 gigabits per second. Cables under development are capable of 5.4 gigabits per second delivery, and 20 gigabits per second cables are being worked on in the research laboratories.

At 2.4 thousand million bits per second, the one and a half hour video could be transmitted in little over a second. So in the course of the 1990s, the standard time taken to send a video to your home will go down from 150 days to a little more than a second!

Sometime in the next century, when the entire telephone network uses all-optical components to build telephone exchanges and switching stations, then an all-optical system will lift the transmission rate by another 1000 times to the 1 000 000 000 000 bits per second level – 1000 gigabits per second or one terabit per second.

These increases in the capability to transmit data will have huge effects on peoples' perception of how to use a telephone line and should be a great boost to anyone selling any product that can be produced in digital form, because delivering that product anywhere in the world should cost virtually nothing.

The hardware costs of doing that are also on the way down because the technology for performing the all-important compression/decompression function can now all be put on a few chips. And once a technology is reduced to chips, it is on the chip price decline curve.

Chips that perform the compression function (and decompression at the other end) were developed in the 1990s and quickly became a big market. As in other areas where chips are involved, prices of compression/decompression chips soon began to fall rapidly, making it a very affordable technology for all.

The effect of compression on video is very much more dramatic than it is on words. The reason why the compression ratio for digitized video is 100:1 when the compression ratio for digitized words is only 3:1 is because it is not necessary to digitize the background of every video frame.

Where backgrounds do not differ much, the background information only has to be digitized once and a symbol used to repeat the information. That makes for considerable savings in storage space.

So the virtually free transmission of data is almost with us. A network of optical cables is gradually being woven into the existing network of copper wires and will gradually replace it. In the USA, and increasingly elsewhere, these optical links are often called 'information superhighways'.

In the same way as motorways got started, the first optical cables/superhighways were laid along the main trunk routes connecting large cities. In the mid-1990s the superhighways are being brought to the high-density population areas and, feeding into and out of the superhighways, will be ordinary wires to carry the information to the home or office. Furthermore, linking into the network will be pocket phones, allowing you to access the network by wireless radio links while you're on the beach or the golf course, in the garden or the hairdresser, in the car or on the move.

As well as the greater accessibility, the main benefit for us all is the plummeting costs of using the telephone. Phone bills are set to tumble over the 1990s – which should stimulate a vast expansion in the way we use the phone service. Obviously telephone-based services will proliferate.

For instance, it will be so much cheaper to link four people together for a sight-and-sound conversation than for them all to travel to the same place to meet up that a great deal of travel will become avoidable.

If the costs of a four- or five-way video-telephone call lasting an hour or two become negligible, then a lot of the time we spend on snarled-up motorways, packed into trains and queuing at over-stretched airports will be avoided simply by linking people via tele-phone wires while they sit in their offices or homes.

With business meetings conducted on multi-way video links, business travel could be considerably cut down. And when you can call up Granny whenever you like on a two-way video link for insignificant cost, even family travel could be cut back.

In the mid-1990s, advances in chip technology will make it possible to add this kind of video-conferencing capability to a personal computer for around $1000. Since this capability will be added in the form of a set of chips, the cost should drop by 30% a year once video-conferencing gets set firmly on the chip industry learning curve and the number of chips required to perform the video-conferencing function keeps reducing.

This could have some dire consequences for the travel business. With the hardware cost of video-conferencing coming down to consumer pricing levels over the rest of this decade, and with negligible costs for telephone calls on the way, the future of the airline, railway and coach travel industries could be pretty flaky.

In time, the cost of communications will be so cheap that the ordinary individual with a camcorder could send digitized holiday videos back home for parents to watch while he or she is still on holiday.

That means everyone will have the ability to be their own TV transmitting station. And, when low-orbit satellites for data communications are in place in the late 1990s, no place on Earth will be too remote from which to send a picture to any other place on Earth.

Meanwhile, the digitization of all forms of electronics activity – TV and radio broadcasting, TV receivers, telephone calls (both wired and wireless), fax machines, computers etc. – is going to have some unexpected results in the look of familiar electronic goods, like TVs, computers, phones and faxes.

For instance, digitization is going to mean that TV sets and computers are going to be pretty much the same thing electronically. That's because both computers and TVs will be dealing with streams of digitized data – the only difference being that the computer gets its stream from a disk or a chip, whereas the TV gets its stream from a broadcast transmission.

When TV broadcasts are all-digital and TV sets are all-digital, then the electronic innards of a TV will look much the same as the electronic innards of a computer. Both a computer and a TV will be capable of handling all kinds of digital information whether it is coming in from a disk, a chip, an audio or video broadcast, a satellite broadcast, a video tape, a telephone line, a wireless telephone or a fax machine.

This could be confusing to the traditional TV set manufacturers and computer manufacturers, who won't know which business they are in! For the microelectronics experts, however, it represents a marvellous opportunity.

The microelectronics industry is exploiting the opportunity by producing chip-sets which can add all these different types of digitized information flow to the computer. These chip-sets start off large and expensive – say six or seven chips costing $2000 – and get cheaper every year as it becomes possible to do the same things on fewer and fewer chips.

First video chip-sets, then telephone chip-sets, fax chip-sets, TV chip-sets, portable telephone chip-sets will all come out, allowing you to hook on these extra information types to the computer.

Eventually, every new function to be added to the computer will get to be done on one chip, and then, whatever that function is, it will be so cheap to add on to a computer that it will become a standard feature on all computers.

The result of this gradual progress towards computers/TVs that do everything with digitized information that it is possible to do is that there will be a merging of all the various machines which at the moment handle their own specific types of information.

Looking back from the first decade of the 21st century we'll be amazed at all the different equipment types we used to have in the 1990s – printers, faxes, answering machines, phones, computers, TVs etc. What will have happened between then and now is that ubiquitous digitization will have reduced them all to a single machine (Fig. 4.2).

The progress towards the all-singing, all-dancing, all-purpose machine will be gradual. For instance, it should be possible from 1995 onwards to buy a chip-set on a card costing about $1000 which will allow a telephone line to be plugged into a personal computer, enabling it to receive video and audio data – effectively turning the computer into a video phone.

The chip-set will halve in number of chips and cost every couple of years until it is down to one or two chips and cheap enough to be included in every computer as a regular feature.

The same will happen for video. First an expensive chip-set will be sold as an additional unit allowing you to plug a video into the computer. Then a cheaper chip-set will be produced and then, when adding video to a computer only takes one or two chips, the feature will be incorporated into computers as a standard feature.

Then again for TV cameras. Chip-sets for plugging these directly into the computer so that the computer can receive transmissions sent from a TV camera or a camcorder will be sold as add-ons. Then these chip-sets will get reduced to a single chip, will be mass-manufactured, made affordable to all and turned into an everyday feature of all computers.

When digital broadcasts become sufficiently widespread, chips will be made that will connect a computer to a TV aerial to receive TV programs.

For functions like faxing, printing or copying, chip-sets are being developed that will connect up the fax machine and the photocopier to the computer, allowing a computer user to send the document on his screen off to a fax machine for transmission or to a photocopier to be copied.

Chips which link up the computer, the fax and the photocopier are due in the second half of the 1990s and will come out as chip-sets on boards that slot into a computer. From there they will rapidly decline in price until it only needs a chip or two to do the job, whereupon they'll be put inside the computer as a standard feature.

The result of all these moves is that the various equipment types we use today will all merge into each other. For instance, two

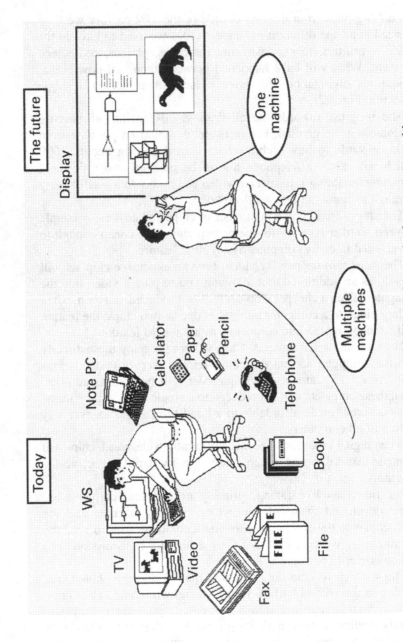

Fig. 4.2 Lifestyle: today – multiple machines; the future – one machine.

machines whose functions look like merging indistinguishably are the fax and the printer.

When printers are connected up to the telephone system they will be able to print out from anywhere on Earth, and when faxes can print on plain paper with letter quality and in colour then there will not be any difference in function between a fax and a printer. These will become a single machine.

And since the single machine will be dealing with a single type of information flow – the 1s and 0s of digital language – it will be able to take on board spoken messages and video messages as well as written messages. So the merged fax/printer adds to itself the functions of the video phone and the telephone.

Meanwhile, pocket telephones will also be taking on more of the functions of other machines – able to send and receive faxes for example. In order to send faxes they will need some of the features of computers such as word processing, and since they will need large sized screens for word processing it will be a short step to add on TV. So the portable telephone becomes a phone/fax/computer/TV.

At the same time, portable computers are expanding to grab the functions of the telephone. First computers got the ability to fit into the telephone network to send messages; now they are getting the ability to send wireless messages to other computers, to fax machines and pagers. After that they will add the ability to show video and then they will be able to receive TV broadcasts. So the computer also becomes a computer/phone/fax/TV.

The result is that a TV becomes like a computer, a computer becomes like a telephone, a fax machine becomes like a printer, and a telephone becomes like a computer/fax/printer/TV.

Eventually everything merges into a single, seamless unit – all brought about by the shrinking transistor continually driving down the cost while increasing the number of functions.

Whether it's a single, seamless unit, a video telephone or a computer with video telephone capability, there is one other function all these will need: the ability to transmit TV pictures as well as see them.

Already the electronics of a TV camera can be compressed onto a single chip, making the electronics cheap. Adding this capability to computers/phones/faxes/TVs or whatever is not a big issue on cost grounds.

The issue is how fast you can send pictures down ordinary telephone wires. The first, early 1990s, videophones only sent pictures at 10 frames per second (whereas video needs 30 frames a second).

Sending pictures at 10 frames per second results in very poor quality pictures and jerky movements.

As we've seen, the transmission rate should be increased substantially in the second half of the 1990s, allowing the transmission of 30-frames-per-second video-quality pictures. Then TV cameras on your phone, computer, TV or fax will be able to send video-quality pictures to another phone/computer/TV/fax.

The progression from separate machines to all-in-one is also partly a question of sheer size. Figure 4.3 shows how the shrinking transistor has affected the actual size of the various equipment products over the years.

To the manufacturers of each specialized machine, the situation is both a puzzle and a challenge. It gives them a chance to muscle in on other markets which they haven't been able to get into, but it means getting into areas about which they know nothing – frequently a recipe for disaster in the electronics industry in the past.

That is probably why, in the mid-1990s, numerous consortia of leading electronics companies from different areas of the industry were formed. Many people thought that the consortia were driven by individual companies' fear of not knowing what to do and ignorance of future technological direction. Gordon Moore of Intel mockingly described the proliferation of collaborations as 'a group grope'.

For the microelectronics industry the future was clearer. The chip people had only to find the most cost-effective way of adding on extra functions to the familiar equipment types – telephone capability to computers and vice versa; fax capability to phones etc.

To do this the chip industry has to provide all the necessary functions while using the fewest number of chips and so at the lowest cost. That's something they have always known how to do. However, if the chip industry lives up to its record, it will be doing more than that.

The industry's history shows that it has always been more than a simple provider of the electronic guts of specific pieces of equipment. The chip industry has always gone beyond that in developing new kinds of chips that change the game.

In these cases, instead of the equipment manufacturers laying down the specifications for chips they want to put in their products, a new chip is invented which tells the equipment makers how they are going to make their products.

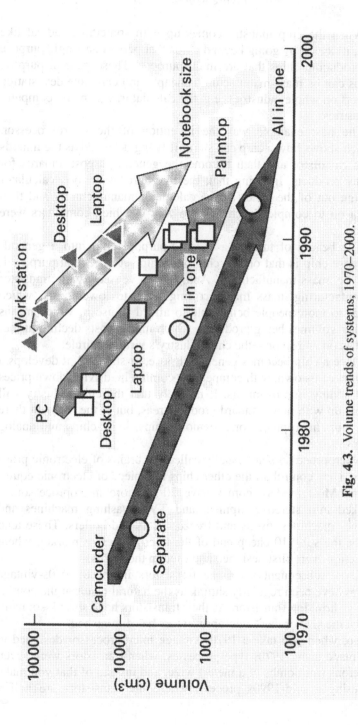

Fig. 4.3. Volume trends of systems, 1970–2000.

When the chip industry comes up with something radical like that, it does it by going beyond chips that perform a single purpose to producing chips that are multi-purpose. These general purpose chips can be mass-manufactured cheaply and can have devastating effects on whole industries, e.g. the calculator, watch and computer industries.

The best example was the invention of the microprocessor, which shows how a chip can start off being designed as the innards of a calculator and then become the general-purpose innards for many products. In doing that, the microprocessor took calculator design out of the hands of the calculator manufacturers and then went on to completely change the way in which computers were made.

The beauty of turning a specific-purpose chip into a general-purpose chip is that once a chip or a chip-set is general purpose it can be mass-manufactured. And, as we've seen with transistor manufacturing, mass manufacturing leads to lower costs, which leads to more people being able to afford to use it, which means higher volumes being produced, which means costs decline further and so on and so on – the chip industry's virtuous circle.

Once a chip becomes general purpose, or standard, it develops a market of its own, with competitors coming in, driving down prices and adding improvements. It is likely that the chip industry will come up with new standard product areas, but at the moment there are four: microprocessors, memory chips, logic chips and analog chips.

Microprocessors are usually called the brains of electronic products. They control all the other chips in a piece of electronic equipment. Most modern homes have 20 or more microprocessors – tucked into stereo equipment and TVs, washing machines and games machines, ovens and freezers, cars and boilers. These tend to be the sub-$10 cheap end of the microprocessor market, where the chips were first designed and used in the 1970s.

Over subsequent years the transistors in these 1970s-vintage chips have been regularly shrunk as the natural result of the continually shrinking transistor. As their transistors have shrunk so, naturally, the size of these microprocessors has diminished.

So, when you take a 1970s-vintage microprocessor designed to be made using 1970s-type processes, when transistors were seven microns (millionths of a meter) wide, and instead of that you make it using a mid-1990s process, where the transistors are half a micron wide – naturally the resulting microprocessor is several times smaller than the original 1970s model.

And because it is so much smaller, very many more can be made from the same piece of silicon. And since the cost of processing the piece of silicon doesn't change much, the more chips squeezed onto the piece of silicon the less each chip costs to make. So 20-year-old microprocessors can be made very small and very inexpensively.

These inexpensive microprocessors are made by many companies, some companies using their own designs, other companies licensing designs. Competition is intense, so prices are low, which means that more people decide to use microprocessors in their products, which in turn increases demand and so pushes up production volumes. Rapidly decreasing cost is one reason why microprocessors are, in the mid-1990s, the fastest growing standard product in the chip market.

The other reason is that, at the top end of the microprocessor market, inhabited by the mighty microprocessors used to power the latest computers, prices are high and, naturally, the higher the price the bigger the monetary value of the market. These top-end microprocessors tend to cost in the hundreds of dollars range, even above $1000, when first introduced.

In this top end of the market a position almost amounting to a monopoly was established by Intel in the 1990s with its microprocessors going into more than 80% of the personal computers being sold worldwide.

The reason why Intel established its dominance was because IBM used an Intel microprocessor in their first PC in 1981, and that became the world standard PC that every other PC-maker wanted to copy. And because the 1981 IBM PC was made from standard off-the-shelf chips that anyone could buy, anyone did buy them.

Lots of clone-makers – companies making machines to run the same software as the IBM PC – were set up, and they dramatically expanded the IBM-compatible PC market. And because this crowd of IBM PC clone-makers all bought their microprocessors from Intel, Intel got big, rich and the dominant player in the microprocessor business. By 1992 it had become the biggest chip company in the world.

Intel's main rival in the microprocessor business, Motorola, fared less well, supplying its microprocessors to the company dominating the remaining 20% of the personal computer market, Apple.

Because the operating software of the Macintosh was not bought off the shelf, and because Apple would not (until a change of policy in 1994) license it to anyone else, no one made clones of the Macintosh. Therefore Motorola's microprocessor did not become such a big seller as Intel's.

However, big, rich and fast-growing monopolies do not usually last long. Potential competitors start knocking at the door – or even kicking it in. That Intel managed to hang onto its dominance for so long was because it stuck to a golden rule which, years before, was established by IBM in the computer business.

The rule is: make sure your new products are compatible with all the old ones, i.e. make sure that every new machine can run every single piece of software ever written for previous machines. By doing that, Intel ensured that its customers never lost the past investment they had made in buying software.

It also meant that a formidable base of software specifically written to run on Intel's microprocessors was built up, so that by the mid-1990s some 50 000 programs worth some $150 billion had been specifically written to run on Intel's microprocessors. And since all this software had been specifically written for Intel microprocessors, it ran faster on Intel's microprocessors than on anyone else's. That helped Intel to stay top dog.

But there was also another threat – from potential rivals who made copies of Intel chips which could run the software as fast as the Intel chips. To see these people off, Intel went to law. Costly lawsuits helped delay rivals' plans, and in addition Intel also did a very clever and unexpected thing.

In the three years 1990–3, Intel spent half a billion dollars on advertising through consumer channels – TV, national newspapers, billboards etc. The product that it advertised was its x86 range (386, 486, Pentium) of microprocessors. The intention of the advertising was to establish a brand name for Intel's microprocessors in the minds of the general public.

No one had ever before thought it possible to establish a brand name for a chip in the public's mind. However, by the mid-1990s it was clear from market research that Intel had established itself as one of the biggest brand names in the American market – on a par with Coca-Cola. It was an astonishing achievement.

However, a far more glittering achievement beckons the successful microprocessor company of the 1990s – domination of the whole computer market, not just the market for personal computers. That's because all computers – not just personal computers – are becoming microprocessor-based. So even big mainframes used in offices and airlines will gradually, as old ones are replaced, come to be made up of banks of microprocessors connected together.

It will take time to replace the existing mainframe computers, but whoever owns the microprocessor market in the New Age of microprocessor-based mainframes will own the computer market. That's why big computer firms like IBM and DEC in the early 1990s started trying to establish own-brand microprocessors.

IBM (in partnership with Apple and Motorola) called its new microprocessor the PowerPC; DEC called its new microprocessor the Alpha. Together IBM and DEC sufficiently rattled Intel for it to announce, in the summer of 1994, that it was teaming up with Hewlett-Packard to co-develop future microprocessors.

Whether the multi-billion dollar challenges to Intel's dominance come off or not will be one of the chip industry's most interesting sagas during the 1990s.

So the microprocessor has grown from humble beginnings – intended to be merely the electronic guts of a calculator it is instead becoming the largest standard product of the chip industry, controlling everything from supercomputers to electric toothbrushes.

The second biggest standard chip product is the memory chip – a chip that simply stores information. There are four kinds of memory chip: dynamic random access memories; static random access memories; erasable programmable read-only memories and flash memories.

Each of them can store information, but each has different characteristics making it more suitable for particular uses. For instance, the memory chip which can always store the most information is the dynamic random access memory – usually just called a DRAM. That's because it only needs one transistor to store one binary digit (a 'bit' for short) – meaning a 1 or a 0.

Therefore a DRAM with 16 million transistors on it can store 16 million 1s and 0s or 16 million bits. Unsurprisingly such a chip is called a 16 megabit DRAM. That's enough bits to store *Gone with the Wind* (the book not the film!). 16 megabit DRAMs cost around $50 in mid-1994 and will most likely decline in price by 30% a year.

Every three years a new generation of DRAM comes out. Each new generation of chip has four times as many transistors as the previous generation and is capable of storing four times as much information.

1994 saw the first emergence of 64 Mbit DRAMs with 64 million transistors capable of storing 64 million bits. They will be two to three years in the evaluation/design-in stage before they are cranked out in high volumes.

DRAMs have the disadvantage of being unable to retain information when the power is turned off. Therefore they are used to store information while it is being worked on or added to. To save the information before the power is turned off, the information that needs storing has to be sent somewhere else – like a disk.

A type of memory chip that does not lose its information when the power is turned off is the erasable programmable read-only memory, or EPROM. Like the DRAM it can store one bit on one transistor, so it can store as much information as a DRAM; however, it is difficult to erase the information. To erase the chip it has to be put under ultraviolet light – so in practice it is not used for working memory.

That's simply because every time an EPROM chip was filled up with information it would have to go through the troublesome process of being exposed to ultraviolet light. Therefore EPROMs are used simply to store standing instructions, like, for instance, the instructions that come up on a computer screen when you first turn it on asking you what options you want to pursue.

A third type of memory chip is called a static random access memory or SRAM. SRAM has a disadvantage against DRAM because it stores four times less information than a DRAM at the same level of process technology (i.e. with the same-sized transistors and line widths), but it is easier to use and can be made much faster.

So SRAMs are often used where it is necessary to have a higher speed memory chip than a DRAM. One purpose for them is to store information that is very frequently used by the microprocessor. If DRAM were used instead of SRAM, the relative slowness of the DRAM would slow down the working of the microprocessor, so reducing the overall performance of the product.

However, SRAMs need four transistors to store a bit and therefore are always three years behind DRAMs and EPROMs in their storage capacity, i.e. when the smallest commercial transistor you can make is half a millionth of a meter across, allowing you to make a 16 million bit DRAM, the highest capacity SRAM it is possible to make will be four million bits.

The fourth main kind of memory is flash. This is a relatively new kind of memory chip. Like a DRAM and an EPROM it stores one bit of information on one transistor, so the latest ones can store 16 million bits. Unlike an EPROM, it is easy to erase information in a flash chip, which means it can be used for working memory. And unlike a DRAM it retains its information even after the power is turned off, so it can be used as the main storage method in equipment.

Flash therefore has terrific potential, because it can be used in every part of a computer – even replacing disks – which is a particularly attractive option because disks need revolving machinery, which needs a lot of power. At the moment flash is used in many mobile phones to store phone numbers and codes, and a growing number of portable computers are using it.

It is not used more because it is a new kind of chip and only reached the 16 million bits per chip capability level in 1993. When 16 Mbit flash is produced in large volumes in the mid-1990s it could be a runaway success.

For flash to be used for both working memory and main memory in computers, it will have to be made as inexpensively as disks. By the beginning of the next century it probably will be. But long before then, all-chip portable computers using flash will be commonplace because people will pay extra for the benefit of having chip memory rather than disks in a portable. The benefit is power saving, allowing for very long battery life.

These four memory chip types – DRAM, SRAM, EPROM and flash – represent the second major segment of the standard chip product area. The third standard product area is that of logic chips, which perform a multiplicity of small tasks inside electronic equipment – adding, subtracting, multiplying dividing etc. – and are produced in vast quantities and get sold, often, for cents.

The standard logic chip segment of the chip market is a stagnant area – not growing and expected to decline. This is because designers of electronic equipment very often find it cost-effective and space saving to bundle up a bunch of the standard logic functions they need and have them all made up together in one specially tailored, or 'customized', chip.

Whether it is worth going to the expense of having customized chips made up, or whether it is easier and cheaper to use standard off-the-shelf chips, is something that varies from time to time depending on various fluctuating industry influences (see Chapter 5).

After microprocessors, memory and logic comes the fourth largest standard product area, which is analog chips. These are chips that do not use the same digital language of the other chips. Instead of using 1s and 0s as their working language they deal in electrical models of natural effects, such as heat, pressure, sound, light and weight.

An example is the following description – incidentally some of the most valuable words ever written: 'vocal or other sounds telegraphically transmitted by causing electrical undulations similar in form to the vibrations of the air accompanying the said vocal or other sounds'. The words come from US Patent No. 174 465, granted in March 1876 to Alexander Graham Bell for his invention of the telephone.

The words describe how Bell based his invention on the principle of the schoolchild's pair of tins connected with a string, which vibrates when you speak into one tin and transmits the vibrations to the other tin, where it can be heard. Bell simulated the vibrations electrically and sent them along wires. Those electrical simulations of the vibrations of a string are an example of an analog signal.

Such simulations of real-world phenomena can be applied to heat and cold, to light and dark, to pressure, to noise, and to weight. Analog chips store and process analog signals and are very commonly used in measuring instruments.

Some people say that digital chips will replace analog ones completely; others say that analog chips are due for a comeback. Some things are done much better by analog chips than by digital chips.

Recognizing shapes and patterns is an example and is an application for analog chips that may be the springboard for its predicted comeback. Many researchers around the world are using analog chip technology to develop means by which machines can recognize shapes – a task at which computers are notoriously bad.

For instance, although digital computers are far superior to a human brain when it comes to calculation, they are far inferior to the brain of an insect when it comes to recognizing a predator. If machines are to become as capable as human brains – and many people presume that is the destiny of machines – then analog chip technology may be the way it is accomplished.

What the machine-brain researchers are doing is trying to create analog chips that are modeled on the neurons of the human brain. These neurons are like millions of tiny microprocessors each one with hundreds of links to the next neuron – so lots of neurons are firing away at the same time. That's what the computer world calls 'parallel processing'.

In the mid-1990s, technology allows neuron-type chips to be built with around a thousand neurons on them. One thousand neurons is the size of the brain of a bee. However it's a start. And as we've seen, once something gets started in the microelectronics business it can develop into something big extraordinarily quickly. For instance, the annual sales of the personal computer industry went from zero to $50 billion in 17 years.

So these are the chips that have brought us virtually free information storage and virtually free information processing and are about to give us virtually free information transmission. They might even give us an artificial brain as good as a human brain. How the chip industry goes about achieving its advances is the subject of the next chapter.

5

How does it happen?

The basic law of the evolution of the microelectronics industry was stated in the 1970s by Gordon Moore of Fairchild, now of Intel (see Fig. 3.1). Moore's Law means that the ability of the microelectronics industry to deliver increasing memory storage and processing capability to the equipment industry is predictable.

For instance, the processing capability of a calculator when it first came out was about 1000 instructions per second (or 1 KIPS). When the personal computer came out it had a processing capability of one million instructions per second (1 MIPS) and when the expected instantaneous speech translator comes out in the year 2000 it will be capable of 1000 million instructions per second (1 GIPS).

It took 15 years to go from 1 KIPS to 1 MIPS and will take another 15 years to go from 1 MIPS to 1 GIPS. In other words, microelectronics capability is increasing by three orders of magnitude every fifteen years.

Memory capacity has been doubling every 18 months (see Fig. 2.1). Every three years a new generation of memory, which has four times the storage capacity of the previous generation, has been put on the market.

How does it happen? Clearly one of the important factors, though only one, is the shrinking transistor. However, there are four other factors of equivalent importance: device invention; package improvements (the way in which the chip is fixed into the end product); architectural innovation and design-by-computer.

The evolution of microelectronics can be looked at as a five-sided pyramid, where each side is represented by one of these five factors. As each side gets to a higher level of capability, so the apex of the pyramid – representing the state of the art of chip technology – gets pushed a little higher.

The first side of the five-sided pyramid is the shrinking transistor itself. The expectation is that it will continue to reduce in size, though not halve, every three years. The 16 million bit DRAMs of 1993 had transistors half a millionth of a meter in size; the 64 million bit DRAMs on the commercial market in 1996 will reduce that to 0.35 of a millionth of a meter.

Impossible, you might think. In order to squeeze four times the transistors onto the same sized chip, it must be necessary to halve the size of the transistors. And 0.35 microns is somewhat bigger than half of 0.5 microns. The answer is that the chips always get slightly bigger.

That's because continuing improvements in the purity of the silicon allow you to use bigger squares of silicon for each chip. Since maybe 200 square chips will be made on a round piece of silicon (called a 'wafer'), the determining factor in the size of each square is the presence (or absence) of defects in the silicon. The fewer defects there are on each wafer, the bigger the square can safely be made without risking a faulty chip.

So increasing silicon purity allows chips to get slightly bigger with each generation, allowing the chip companies to quadruple the transistor count without actually having to halve the transistor size with every generation.

Together, the effect of smaller transistors and larger chip sizes combine to give us the four times increase in capability every three years. The transistor size reduction contributes about two thirds of this increase in capability and the larger chip area contributes the other one third.

That is to say the transistor size reduces by 60% with each generation, providing a density increase of 2.8 times and the chip area increases 1.4 times each generation, contributing the rest of the quadrupling in the number of transistors.

So 256 megabit chips, expected on the commercial market in 1999, will have transistors of 0.2–0.25 microns in size and the 1 gigabit chips marketed in 2002 will have 0.1–0.15 micron transistors.

Laboratory examples of chips are made years before they appear on the market. Prototype 256 megabit chips were in existence in laboratories in 1993, some six years before they are expected to be seen on the market.

At the same time, old memory chips are staying on the market for longer and longer. For instance, in 1994 the first 64 megabit chips appeared while the 16 megabit was in volume production and the chip being made in the highest volume was the 4 megabit. Even so the 1 megabit and the 256 kilobit were still being made and sold and used.

That means five generations of microelectronics technology – representing 15 years of technological progress – are all on the market at the same time. This is a phenomenon that may extend as chip factories get more and more expensive to build.

Not only does the cost of chip factories increase dramatically with each new generation, but, according to the American industry analysts Dataquest, the increase is rising more steeply than in the past. From a cost of around $60 million for the 64 kilobit chip, it rose to around $180 million for the 256 kilobit, to around $300 million for the 1 megabit chip, to around $425 million for the 4 megabit, to around $700 million for the 16 megabit, to $1 billion for the 64 megabit, $1.4 billion for the 256 megabit and to $1.9 billion for the 1 gigabit chip.

The enormous costs come mostly from the production machinery and the cost of filtering out impurities in the air. Other important considerations are ultra-pure water, chemicals and gases and minimal vibration.

In the case of the 1000 million bit (gigabit) memory chip which is expected to be first made in prototype by the end of this decade, the transistors are expected to measure around 0.1 millionths of a meter in size and the whole chip is expected to be around 20 millimeters square.

In order to make the 1 gigabit chip efficiently, it will have to be made in a factory where every particle of dust larger than 0.1 millionth of a meter has to be extracted. (Otherwise the dust landing on the chips would create short circuits in the chips and most would be duds).

To give an example of the problems involved, it is useful to compare a chip to a football field (Fig. 5.1). If you extrapolate the chip size to the size of a football field, then the 0.1 millionth of a meter transistor size corresponds to an object measuring half a millimeter on the football field. So the task in making a factory clean enough to make 1 gigabit chips efficiently corresponds to removing all particles bigger than half a millimeter from a football field.

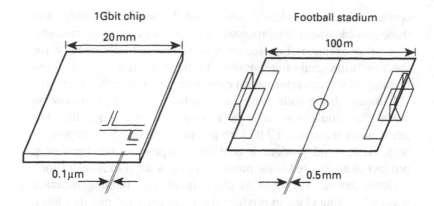

Fig. 5.1 How narrow is the line width on 1 gigabit chip?

The cost of building a factory to do that is expected to approach $2 billion. These vast sums to make the latest chips can only be justified if the factories keep on making older chips long after they are overtaken by new ones.

Moreover, the horrific expense of making new chips could mean that the cost per memory bit, which naturally fell by four times with every succeeding generation, may only fall to half that of the previous generation.

So what is likely to happen is that prices will not drop as quickly as they used to – which used to force people to use the new generation of chips. Instead it will not be so cost-effective to move to new chips and the older chips are likely to stick around for longer in the future.

Since these older chips will be being made in fully depreciated factories, and since the R&D costs will have been paid off, they should be able to be made inexpensively, which will be a great benefit to all those who do not need the latest technology.

So much for the contribution that the process of shrinking the transistors makes to raising the apex of the five-sided technology pyramid; the second side to the pyramid is the invention of new types of chip, or device invention.

It is device invention which tends to produce the epoch-making products. For instance, a transistor radio was developed when the germanium transistor was invented, the electronic watch was made possible when a special kind of chip called a complementary metal oxide semiconductor (CMOS) chip was invented, which could operate off very little electric power, and the personal computer was made possible when microprocessors and memory chips were invented.

Another example of device invention is flash. Flash memory, for the first time, makes it possible to have a high capacity chip memory that stores information even when the power is turned off.

It means that a wide range of portable equipment can now be made that could not have been made before. And, like the germanium transistor, CMOS chips, memories and microprocessors, flash could produce a complete surprise in the form of a product as unexpected as the transistor radio or the pocket calculator.

Other device inventions in chips involve combining different ways of making chips to produce the ideal chip, i.e. one that has a lot of transistors, works at high speed and uses very little power.

The combinations involve taking the best features the various technologies used in chip-making and trying to put them together to make a superchip. The main technologies are: bipolar technology, which offers high speed; n-channel metal oxide semiconductor (NMOS) technology, which offers high transistor count; and complementary metal oxide semiconductor (CMOS) technology, which offers low power requirement.

Much of today's laboratory work at chip companies involves tinkering with combinations of these technologies to try to produce the perfect blend for the superchip of the future. It should offer an amalgam of low power, high speed and lots of transistors. An example of the sort of 'dream chip' that such a combination might produce is a memory chip with a large storage capacity that keeps its memory without power and works at high speed.

Flash provides the first two but not the third. The industry name for this ideal dream chip is a 'non-volatile RAM' – non-volatile meaning that it doesn't lose its memory when the power is turned off. In the mid-1990s, there are hopes among chip-makers that a technology called ferroelectric memory may be the route to the dream chip.

The move towards low power comes, to a degree, as a natural function of the industry as it moves to lower and lower power simply because the smaller the transistors get the less power they need to switch on and off.

At the half millionth of a meter size – reached in the mid-1990s – the standard voltage needed for chips dropped from five volts to three volts, and by the end of the decade – when transistor size is between 0.1 and 0.15 millionth of a meter – the standard is expected to drop again to two volts.

Another way to get to better performing devices is to use a different type of material from silicon. For years a material called gallium arsenide has been touted as 'the technology of the future' (to which the sceptics say 'and always will be').

The promise of gallium arsenide is that electrons move through it six times faster than they do through silicon – so offering very high performance. However, the difficulty in using gallium arsenide has always meant that it has turned out to be much more expensive than silicon.

A more promising route could be the amalgamation of germanium and silicon in an alloy. The material allows for very high speeds at very low power and is relatively inexpensive. Experimental chips based on silicon–germanium first appeared in the mid-1990s.

So the second side of the five-sided chip pyramid is device invention. The invention refers either to improvements in the material or to improvements in combining the best bits of different technologies or to improvements in simply coming up with a new kind of chip that pushes the apex of achievement of the chip pyramid a bit higher.

The third side of the pyramid is how you fit the chip into an electronics product. Obviously, a thin, delicately patterned, fingernail-sized piece of material is not much use on its own – it has to be connected up to the outside world.

Ways are continually being found to make the frame which holds the chip – known as the 'package' – smaller and smaller. At the same time, the chip size, as we have seen, is getting bigger and bigger. Just as you cannot wear shoes smaller than your feet, you can't have a package smaller than the chip.

Nowadays the package has been reduced in size and shape until it is only fractionally larger than the silicon (Fig. 5.2). For the future it may be necessary to go vertical – like stacking up chips on top of each other to make 3D structures.

The fourth side to the pyramid is innovation within the chip itself – generally referred to in the chip business as 'architectural' innovation. An example of architectural innovation was the addition of memory on the microprocessor chip to store those instructions that are most commonly used by the microprocessor. Having those instructions close to the processing kernel of the chip speeds up the rate at which the microprocessor operates.

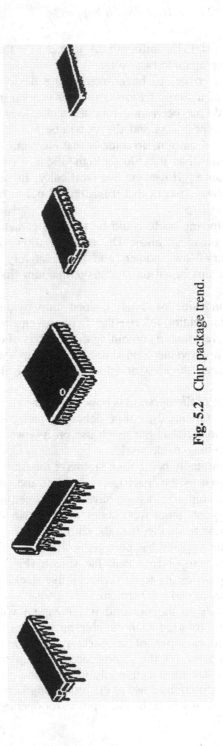

Fig. 5.2 Chip package trend.

Another example of evolutionary architectural innovation is the size of the chunk of information that the microprocessor can handle at any one time. The first microprocessors handled four-bit chunks, i.e. they could handle four bits of information at once (four bits being four binary digits – either 1s or 0s).

These 4-bit microprocessors, as they were known, were soon followed by microprocessors handling 8 bits of information at a time and then 16 bits. For the first half of the 1990s the standard was 32 bits, but in the second half it will be 64 bits.

Another big architectural advance was getting microprocessors to perform more than one instruction per cycle. A microprocessor's speed is measured in cycles per second, and until an architectural advance called 'super-scalar' came along, microprocessors could only perform one instruction per cycle.

Devised in the 1980s, the super-scalar architecture allowed the microprocessors of the early 1990s to perform two instructions per cycle and sometimes three, with four in prospect.

Super-scalar architectures were one of a number of significant 1980s advances in microprocessor architecture developed by the American universities of U.C. Berkeley and Stanford and by IBM that were given the generic name of RISC (reduced instruction set computing). They aimed to replace the existing CISC (complex instruction set computing) microprocessors with new architectural features that would speed up the workings of microprocessors.

As the name suggests, the key was to cut down the number of instructions stored by the microprocessor to only those that were most commonly used and store them in a block of memory transistors on the chip itself. This speeds up the operation of the microprocessor because it cuts down the number of times it has to go to outside chips for instructions – a much slower process than going to an on-chip memory bank.

Super-scalar and reduced instruction sets are two of a number of features, generically named RISC, which have been adopted by the designers of most modern mainstream microprocessors. At one time it was seen as a RISC vs. CISC battle with one type or the other type dominating. However, in the 1990s the differences are blurred, as the traditional CISC designers adopt RISC features.

The arrival of the RISC techniques resulted in dramatic increases in the capabilities of microprocessors. For instance, by using super-scalar architecture, which allows you to process more than one instruction per cycle, you increase the throughput of your microprocessor without having to increase its speed.

And since the speed of microprocessors has been increasing exponentially in the last ten years or so, the rate of performance increase in microprocessors has been dramatic – by over 100 times between 1982 and 1994 – with no end in sight.

The fifth dynamic that is pushing chip technology forward – the final side to the chip pyramid – is the use of computers to design chips. From designing chips by hand in the 1960s, the chip industry has gradually automated the process of designing chips.

As chips have rapidly become more complicated, the task of designing them would have taken far too long and needed more designers than existed in the world had computers not taken over the task.

Chip design at the largest companies is done using supercomputers, so complicated is the task. However, the human element remains just as necessary, and the design teams for very complicated new chips like microprocessors can exceed a hundred people, arranged in groups to deal with different parts of the chip.

Although it is possible to envisage the ultimate computer for designing chips, which will be able to take a description of the desired functions of the chip and create the design from the description, that is only possible where all the various parts of the chip have been designed before. Where new ideas are being incorporated in the chip only a human will do.

Although these five elements – shrinking transistors, device invention, packaging, architectural innovation, and design-by-computer – all push forward the state of the art in chip technology, they are not always all pushing forward at the same pace.

Sometimes there will be great leaps in the techniques of chip design by computer, or sometimes there will be a big architectural advance such as the invention of the microprocessor. When that happens, the result is a shift in the focus of the chip industry.

For instance when there are great advances in design-by-computer, it becomes cost-effective for equipment manufacturers to have chips made specifically for them: in effect, tailor-made chips.

When, however, architectural innovation is strong – e.g. the microprocessor is invented – then it becomes more cost effective for the equipment manufacturers to buy a standard off-the-shelf part and program it themselves to suit their own purposes.

So the industry tends to shift from periods of customization – when the emphasis is on tailor-made chips, to periods of standardization – when the emphasis is on standard products.

Historically these periods can be tracked, and they slip into ten-year cycles. One of the co-authors of this book devised a model to demonstrate the cycles which was cited in *IEEE Spectrum*, the journal of the American Institute of Electrical and Electronics Engineers and in the UK electronics magazine *Electronics Weekly*, under the tag 'Makimoto's Wave' (Fig. 5.3).

The balance that the chip industry has always had to strike is between pleasing the customer, which means giving the equipment manufacturers exactly what they want – usually by tailor-made chips – and the operational needs of the chip-makers, who want to be able to turn out high volumes of products in long production runs for maximum efficiency of operation.

If a microelectronics company veers too far towards custom products, it loses the operational efficiency of long production runs; however, if it goes too far towards standard products, then it runs the risk of not satisfying customers. So, after a period of veering one way, the industry has swung back the other way, and these periods of customization and standardization have alternated every ten years.

The periods of standardization have followed on the invention of some epoch-making device but there has always been a period of incubation – about ten years usually – from the time the epochal product was invented to the time when it was a high-volume standard product forming the basis of a standardization cycle.

For instance the 1957–67 decade of standardization followed the invention of the transistor (1947), which became commercially manufacturable as a result of the invention of the germanium junction transistor (1950), which became widely used when the silicon transistor was invented (1954), and which was manufactured in very large volumes when production yields of transistors were dramatically increased by the invention of the mesa transistor (1957).

During the 1957–67 decade, many manufacturers made transistors that were interchangeable in their end use, ushering in an era of standard product domination. As the 1960s progressed, so the number of transistor companies increased, competition intensified, prices plummeted and there was the first microelectronics industry crash. It was the first of many.

A decade of customization followed. The epoch-making development that had been incubating for ten years before the 1967–77 period of customization was Jack Kilby's chip, developed in 1958.

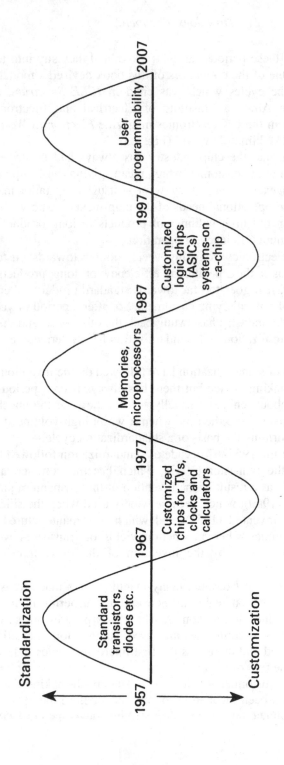

Fig. 5.3 Makimoto's Wave.

The chip ushered in an era of customization. It was first mainly used in space as a tailored device performing functions stated in the specification of the equipment builder – very much a customized product.

The same applied to the chip's second biggest user – the calculator manufacturers, who told the chip companies what they wanted the chips to do and the chip-makers then made them that way.

Others taking up the chip were the clock- and watch-makers. The chip allowed TV-makers to make sub-systems-on-a-chip and the clock- and watch-makers commissioned the chip companies to make customized chips of ever-increasing complexity as more and more features were added to the products to make them competitive.

It was therefore no surprise that, in the 1967–77 decade of customization, the first chip company to make a success entirely out of the custom chip business emerged – American Microsystems Inc (AMI).

However, incubating away during this decade of standardization were a couple of inventions that were going to end it. Memory chips had been invented in 1969 and the microprocessor in 1971.

The invention of the microprocessor was the immediate catalyst for the change. As we've seen, some of the biggest users of the chip during this decade were the calculator companies. More and more of these set up and the industry became very competitive. In order to get ahead of rivals, calculator companies asked the chip-makers for more and more special features.

As more and more calculator manufacturers put on more and more features, the product lifetime of calculators got shorter and shorter as they became superseded by newer, better models with more features. This meant the chip-makers had to keep making new chips with shorter and shorter life-cycles, which is the most inefficient way of running a chip plant there is.

Then Ted Hoff, of Intel, working with the Japanese calculator company Busicon, came up with the brilliant notion of making a set of chips that could be upgraded to add new features by the calculator companies themselves as and when they wanted.

It meant that the chip companies could keep on churning out the same old chip in long production runs, which maintained its operational efficiency, and the calculator guys could keep adding on new features to their calculators whenever they saw their competition getting to be too good.

It seemed the perfect answer to the problems of both the chip companies and the calculator companies. The chip companies could manufacture in volume – which suits the high fixed overhead nature of the chip business – and the calculator companies could add features to their products without having to go through the lengthy and expensive business of having an entire new chip-set engineered each time.

But Hoff had done more than produce a set of chips that was only useful in calculators: it was useful in almost every type of electronic product. Which is why, after six years' incubation following Hoff's invention, the microprocessor ushered in the 1977–87 decade of standardization.

The story is a good illustration of the forces that push the chip industry towards customization or standardization – an extreme move in one direction provokes a counter-reaction pushing the industry towards the other direction.

The other epoch-making chip incubating during the 1970s was the memory chip. By the start of the 1977–87 decade of standardization, memory chips accounted for sales of $800 million annually, but by the end of the standardization decade that figure had soared to $6 billion annually – an annual growth figure of 34%.

At the same time, microprocessor sales were rocketing. During the 1977–87 decade microprocessors went from being sold in quantities of a few million units in 1977 to 750 million units in 1987.

Driving growth throughout the decade were the Apple II, launched in 1978, and the IBM PC, launched in 1981. And naturally, as we saw with the transistor manufacturers, everyone piled in to make competing products.

At that time, the early 1980s, all the major Japanese chip companies as well as all the major American chip companies were making memory chips and, just before Christmas 1983, a traumatic event for the Americans occurred. Boosted by government chip development programs two Japanese companies – NEC and Hitachi – began full-scale production of a 256 kilobit DRAM a full year before the Americans were able to produce it.

For the Americans, who had invented the chip industry and who regarded their technological pre-eminence almost as a divine right, it was a sobering experience.

But worse was to come. Throughout 1983 and most of 1984 demand for memories seemed unstoppable. Everyone in the industry added factory capacity and new people and cranked up volumes until in the second half of 1984 – demand suddenly slumped. That left the chip-makers in a position of gross oversupply of memory chips, which led to catastrophic price erosion with manufacturers selling memories for any price they could get.

In 1985 the world chip industry lost an estimated $6 billion. While all but one of the American manufacturers pulled out of making DRAMs – the most technically advanced memory chip requiring the most R&D dollars – every single Japanese producer stayed making DRAMs.

So that was how the great standardization decade of 1977–87 came to its sorry end. It was, naturally, succeeded by a decade of customization. This decade was based on a concept for chips called 'applications specific integrated circuits' or ASICs.

The basic ASIC product was the MOS gate array which had been incubating since 1977. It was a chip that was partly standard, in that it could be made in its base form as a standard volume product, but it could be tailored to a customer's particular purpose by a final production process. As such, the gate array allowed customers to have their own designs put into silicon relatively cheaply.

Adding to the potential offered by the gate array were advances in chip design by computer which made it much simpler and quicker to design chips. Some equipment manufacturers could use computers to design their own chips without having to go to the chip companies. This made them very much more inclined to use custom chips because they could keep their design secrets to themselves. It also cut costs to use their own designers.

Meanwhile, the chip-makers formed 'libraries' of old chip products that customers could order up and have included in their designs. In this way several separate chips, reduced in size by generations of transistor shrinking, could be put together on one chip to form an electronic system or sub-system on a single piece of silicon.

It became a popular way of making chip-sets for complicated apparatus such as camcorders or products involving radio communications. Such products need a variety of different technologies – say analog chips as well as digital chips combined together on one chip – and gate arrays were the lowest-cost method of putting a number of different chip types together on one chip.

Gate arrays gradually took over the place of standard logic chips in many products, as the functions of a number of separate standard logic chips could all be incorporated together on one gate array chip. By 1987, the ASIC market was already larger than the standard logic market.

So, taking the start of the microelectronics era as the invention of the transistor, it is possible to divide the industry's progress into five decades:

1947–57: the decade of R&D

1957–67: the start of commercialization of transistor production, leading to new products like portable radios and TVs

1967–77: the start of commercialization of the chip, leading to new products such as the calculator

1977–87: the start of the commercial take-off of memory chips and microprocessors, leading to new products, such as the personal computer

1987–97: the decade of ASIC, providing specific chips tailored to specific products, like camcorders and notebook computers

The transistor decade was standard-product-oriented, the chip decade was custom-oriented, the microprocessor/memory decade was standard-product-oriented and the ASIC decade was custom-oriented.

Each decade came to an end when either the product runs of the custom decade became too short for the chip-maker to be able to operate efficiently or when the rampant competition of the standard product decade caused oversupply, with prices dropping to uneconomic levels.

After the ASIC decade of customization comes a new decade of standardization, starting in 1997. The product that has been incubating to make it happen is a new type of logic chip, which is programmed by the user.

A number of small American companies – Altera, Lattice, Xilinx, Actel and others – have pioneered the concept and in the mid-1990s started to become a significant collective force in the chip market after incubating their products for around ten years.

The difference between the new concept of programmable logic and ASIC is that with ASIC customers give their requirements for a chip, or their design of a chip, to a chip-maker, who makes it into silicon. With the new programmable logic chips, these are churned out as standard products by the chip-makers and are then programmed to their specific use by the customers themselves. In some cases they can be wiped clean of their programming and used again for a completely different purpose.

The beauty of the new programmable chips is that they can be turned out in volume as a standard product – so attracting the benefits of high-volume manufacturing and potential for cost-cutting – but, at the same time, the chips can be tailored to fit a customer's precise requirement.

And because programmable logic chips can be bought in small quantities, it doesn't matter if the product into which they are designed is only going to be made in small quantities, or if the lifetime of the product is going to be short. And the early to mid-1990s were a time when product lifetimes and equipment production runs were short and getting shorter.

By the mid-1990s programmable logic technology became mature enough and the market looked large enough for programmable chips to become the basis of a new decade of standardization. The market size of the programmable logic business in the mid-1990s approached $1 billion – the same size as the 1977 market for microprocessors and memories at the beginning of the standard product decade of 1977–87.

So programmable logic is one new kind of standard programmable part – just as the microprocessor was a standard programmable part – that could drive the new wave of standardization coming along after 1997.

The other force that could drive the next standardization cycle is the emergence of flash – a form of cheap, easily reprogrammable memory that retains its contents when the power is turned off. It gets its low cost from only needing one transistor to store one information bit.

By adding flash memory to a great range of chips it will be possible for equipment-makers to program and reprogram them at their will. This will give people the potential to develop new standard chips, where part of the chip is 'fixed' – in that its functions are unchangeable – but another part, the flash part, is programmable by the customer.

That means the chip-makers can come up with more varieties of standard chips made in high volumes and, by adding flash to them, customers will be able to add on their own bit of programming to give them the special feature or extra competitive edge which they want. This in turn means that lots of different customers can use them for many different purposes, so providing the chip industry's virtuous circle of more customers = higher volumes = cheaper prices = more customers.

So flash memory and programmable logic could be the harbingers of the 1997–2007 decade of standardization. And a decade of standardization is often a time of innovation for the equipment industry, as it finds new things to do with the new products thrown up by the chip people.

And a decade of standardization is usually an expansive time for the chip industry, because manufacturing standard products is more in tune with the underlying economics of the chip business, which work best when churning out standard products in high volumes.

However, quite apart from the cycles of standardization and customization, and acting quite independently, is the notorious Silicon Cycle. The chip market has always fluctuated wildly – lurching from 50% market growth in a year to 20% market decline in a year, sometimes – catastrophically – in succeeding years.

Handling those fluctuations has been a major problem for the industry's managers. However a rule of thumb can be used as a guide and it is reproduced here (Fig. 5.4).

The silicon cycle is a four-year cycle and until 1988 its peaks coincided with the holding of the Olympic Games. Various theories have been proposed about why this should have been so – such as people buying more TV sets when the Olympic Games happen. However, the Barcelona Games showed that all the theories were wrong – 1992 was a disastrous year for the chip industry. The following year was a particularly good one.

So the industry's ups and downs are unpredictable, although the overall direction for the industry has been steeply upwards. Averaging out the annual growth of the microelectronics industry for the 25 years from 1969 to 1994 you get a compound annual growth of 16.5%.

According to estimates by Texas Instruments, the chip industry should be doing around $200 billion worth of business annually by the end of the 1990s and that implies a value for the electronics industry of the year 2000 of around $1 trillion – since the value of the microelectronics content of electronics equipment products is expected to rise to 20% by 2000.

At $1 trillion, the electronics industry of 2000 will probably be the largest industry in the world, ahead of steel, cars, chemicals and pharmaceuticals.

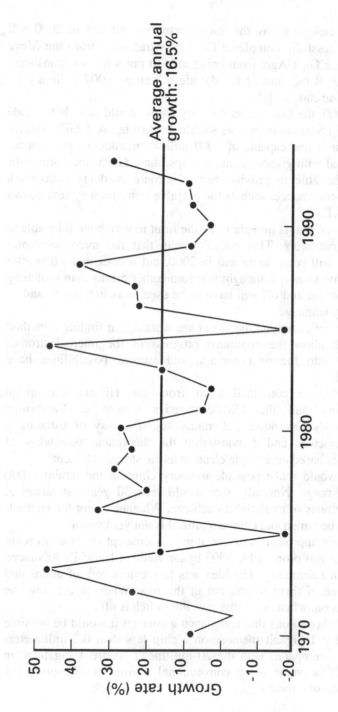

Fig. 5.4 The Silicon Cycle, 1970–94.

In technology terms, the microelectronics industry of 2000 will have successfully completed the 1990s transition from the Mega Age to the Giga Age: from being able to put a million transistors on a chip at the start of the decade to putting 1000 million on a chip by the end.

By 2000, the first 1 Gbit memory chips should have been made and 1 GIPS microprocessors should be routine. A 1 GIPS microprocessor is one capable of 1000 million instructions per second. Combined with gigabit memory chips, the 1 GIPS microprocessor should be able to provide new electronic products with much higher performance, such as the portable translation system shown in Fig. 4.1.

However, that is thought to be the limit to which we'll be able to shrink transistors. That doesn't mean that the microelectronics industry will come to an end in 2000, but it does mean that after 0.1 micron sizes it is thought that some other means than switching transistors on and off will have to be used to switch the 1s and 0s of binary language.

Laboratories around the world are working on finding a method that will allow the continued progress of the microelectronics industry into the next century, and various possibilities have emerged.

In 1993, a combined team from the Hitachi Cambridge Laboratory and the Microelectronics Centre of Cambridge University's Cavendish Laboratory found a way of isolating a single electron and demonstrated the theoretical possibility of memories based on a single electron as the storage element.

This would make possible memory chips in the terabit (1000 gigabit) range. Normally that would take 30 years at historical industry rates of evolution to achieve. Whether or not the technology can be tamed and commercialized is not yet known.

Another approach is to use atoms. A concept of an atom relay transistor was proposed in 1993 by Dr Wada of Hitachi's Advanced Research Laboratory. The idea was to create a row of atoms and move one of them in and out of the row. When in the row, the switch is on; when out of the row, the switch is off.

Dr Wada reckons that using such a concept it would be possible to make a 1 gigabit memory on a chip less than 0.2 millimeters square – compared with the 20 millimeter square 1 gigabit chip that will be made using conventional transistors measuring 0.1 millionths of a meter.

As with the single-electron memory, the atom relay transistor is a laboratory idea, and commercial exploitation – if it ever happens – is ten to twenty years away.

Another alternative is superconductivity – using materials that have little electrical resistance – which allows very fast movement of electrons through the material and so very high performance. Despite great excitement at the end of the 1980s, no commercially useful superconducting material for chips had been developed by the mid-1990s. The only materials that had been developed by then were superconducting only when deep-frozen, which is not a practical option in the commercial chip market.

Another prospect is to use plastic molecules as switches. It has been shown that molecules in plastic can conduct electricity, and it is therefore possible to make a plastic-based On/Off switch that could be the basis of plastic-based chips and computers.

Using light for switching is another possibility, but it is an unlikely contender in the mainstream chip business because of the huge investment that would have to be made in the technology to make it useful.

No matter how it will be done, one thing seems certain: people will continue to want more and more memory and more and more processing power. Today, with a world population of 5.6 billion people, each person uses 800 000 bits of computer memory on average.

By the end of the decade, when the world population is expected to be 6 billion, each person will be using an average 8 million bits – the storage capacity of a 250 page book.

For companies that can keep on the technology treadmill, making ever-smaller transistors, and which can read the industry standardization/customization cycles right and which don't get caught by the notorious four-year Silicon Cycle, the rewards will be rich. Some of those sharing in them will be found in Chapter 6, though, if the past is any guide, not all of them.

6

Who are the players?

Surprisingly few companies develop, manufacture and market microelectronics technology – fewer than 200 worldwide – though if you count in design houses, which offer chip design services to chip-makers and chip-users, the number is larger.

Maybe the reason why there are so few is that the entry-level qualification requires two commodities that are relatively scarce – large amounts of money and intellect.

The financial requirement for R&D and factories is around a billion dollars each in the mid-1990s. $1 billion will buy either a new factory or pay for the R&D for a new generation of memory chips or a new generation of microprocessor. And if you're a major player, you'll need both the R&D and the factory.

This means that the microelectronics industry is dominated by its major players. The ten biggest companies in the world usually account for about half the value of the chips sold in the world.

That means that it is important to be in, or near, the top ten, because only sizable revenues allow you to keep investing in the all-important technology shifts that keep coming along. However, you also have to bet on the right technology shift; if you bet on the wrong direction, or if you invest too little or too late, then you lose your revenues, which means you have less to invest in future technology development, which in turn leads to further diminution of revenues and a downsizing of the company.

So chip companies tend to be either on a virtuous circle upwards of greater investment leading to more revenues or a vicious circle down of less investment leading to diminished revenues. Like the old adage about boxing champions, chip companies never come back.

Although it is important to get big quickly and stay big, so volatile are the fortunes of companies in an industry where the technology shifts are dramatic and unpredictable that the top ten list of players tends to change regularly and radically.

For instance, only three of the top ten tube-makers, General Electric, Sylvania and RCA, made it into the top ten of the transistor-makers. And if you take the top ten of the transistor-makers in 1955, only six survived to be in the top ten microelectronics manufacturers of 1965, when the chip market was getting under way. The six were Texas Instruments, Motorola, General Electric, RCA, Transitron and Philco. Elbowing their way in were the newcomers Fairchild, General Instruments, Sprague and Raytheon.

Ten years on, in 1975, the fallers were Sprague, General Electric, Transitron, Raytheon and Philco – they had been elbowed out by newcomers National Semiconductor, Intel, Rockwell, Signetics and AMI. By 1985, AMI, General Instruments and Rockwell had been replaced by Harris, Mostek and Advanced Micro Devices (AMD).

Throughout the period 1960–87, Texas Instruments, the chip's co-inventor, maintained the remarkable record of keeping the No. 1 slot. The only other three companies to stay in the top ten all that time were Motorola, Fairchild and RCA. The other remarkable thing is that, in the 30 years from 1955–85 – from the tube, through the transistor to the chip – RCA remained a top ten player.

It is also notable that the Americans dominated the industry they invented until the late 1980s. The crunch came in the watershed year of 1985, when the worldwide chip industry lost $6 billion and all but one of the American companies pulled out of the largest product area – DRAMs – while all the Japanese companies stayed in.

The effect on the top ten rankings showed up three years later when, in 1988, six US companies dropped out of the world top ten, replaced by six Japanese companies.

The other traumatic event for the Americans was that they lost the coveted No. 1 spot. Texas Instruments' long reign as No. 1 had ended by 1988 – replaced by NEC – though another American company, Intel, regained the championship for the USA in 1992.

For the Europeans, its one and only top ten company Philips dropped out in 1993 to be replaced by the first-ever company to come from outside the USA/Japan/Europe triad – Samsung of South Korea.

The fluctuations of the top ten are matched by equally severe fluctuations among the lesser ranks. Companies tend to start with one great idea, exploit it and grow big – and then fail to have a second great idea. They wither and die as the first idea gets overtaken by the great ideas of new companies.

Any look at the players in the microelectronics industry should begin with the inventors of the tube, the transistor and the chip. After dealing with those, the next 25 companies mentioned are in order of size of 1993 turnover according to the US research company Dataquest. The rest are more or less randomly selected on grounds of general interest.

GENERAL ELECTRIC

The genesis of the tube was in the work of Thomas Edison, and the heir of Edison was General Electric of America (GE), which bought out the commercial interests of Edison's 'invention factory'.

GE was one of the leading tube-makers and made a successful transition to the age of the transistor. It was No. 3 in tubes in 1955 and was the world's No. 5 in transistors in 1965. GE also made the transition to the chip era successfully, remaining a top ten player in microelectronics until 1975. Thereafter the company languished as a second-tier US player in the industry. In the 1980s, under a corporate policy of exiting any business in which the company was not No. 1 or No. 2 in the world, GE sold its microelectronics business to Harris Corporation.

AT&T

The inventors of the transistor, AT&T, evolved from the commercial operations of Alexander Graham Bell, inventor of the telephone. The company is still called 'Ma Bell'. Its invention of the transistor in 1947 and subsequent dissemination of the technology to the worldwide electronics industry via the Bell Symposium of 1952 were not for altruistic reasons, but to head off mounting criticisms of its role as owner of the US telephone network and so the monopoly supplier of telephone services to America.

This lucrative monopoly allowed AT&T to fund the largest research establishment in the world – Bell laboratories – which employed some 6000 people at the end of the 1940s. As well as the original transistor, Bell Labs made many contributions including the more manufacturable 'mesa' transistor and development of efficient transistor-manufacturing processes.

However, the company could not ward off attacks on its monopoly forever, and in the 1980s deregulation of the telecommunications industry meant AT&T had to become more of a commercial player. After a longish period of adjustment its chip-making arm emerged as a second-tier player in the microelectronics industry.

FAIRCHILD

A chip company that is no longer with us but which made a great – some would say the greatest – contribution to advancing micro-electronics technology is Fairchild Semiconductor.

Formed by the 'treacherous eight' ex-Shockleyites, as Shockley himself dubbed them, so rich was the technology legacy of Fairchild that the company lived on its royalties derived from patents for years after the leading lights had left.

In 1979, however, Fairchild was bought by the French oil-field services company Schlumberger. In 1987 Schlumberger sold it to National Semiconductor, and the Fairchild name disappeared.

Fairchild both invented and defined the chip industry. It co-invented its product – the chip – and it invented the production process by which the product could be mass-produced. It also defined how the industry should be run – by engineers, without hierarchy or status symbols or offices. And it pioneered that most potent vehicle for microelectronics innovation – the Silicon Valley start-up company.

TEXAS INSTRUMENTS

Texas Instruments had already made a significant contribution to microelectronics before Jack Kilby devised his chip.

It invented the silicon transistor in 1954 when the entire industry thought it was impossible. The leader of the team that developed it, Gordon Teal, had been recruited from Bell Labs in 1953 after Texas had attended the 1952 Bell Symposium to acquire transistor technology. Until that time, Texas had made instruments for oil exploration.

Texas was smart enough to really push the progress of the silicon transistor from invention to the commercial market-place – getting out something that worked rather than something perfect. So accelerated was the silicon transistor's development that it was put on the market the same year its invention was announced.

The rush to market had two advantages – it by-passed the purist researchers who wanted perfection and it gave Texas a very lucrative three-year monopoly in the market – particularly profitable because of the interest of the US military.

Then in the summer of 1958, a new recruit, Jack Kilby, earned his place in history with the development of the chip – TI's greatest contribution to the industry.

Texas has retained close links with the US military and receives massive military support for its R&D. In the commercial marketplace Texas has made its revenues from DRAMs and particularly from its development, in the 1970s, of a range of standard logic chips made in bipolar technology known as low-power Schottky or '74 LS'. The 74 series of logic functions was widely copied and remained the backbone of Texas' revenues until the mid-1980s, when MOS technology took over.

Texas' 25 years as the world's No. 1 chip company will probably never be equalled.

RCA

Another of the great pioneers of the microelectronics industry, though now out of the game, was RCA.

The Radio Corporation of America (RCA) and GE of the US were the only companies to accomplish successfully the feat of being major players in three generations of technology: the tube, the transistor and the chip.

In 1955 RCA was both the top tube-maker in the world and the seventh largest transistor-maker. Thirty years later it was the eighth largest chip-maker in the world. That is a tribute to its management who successfully adapted to change when others stuck too long with outdated technologies or failed to bet correctly on future ones.

RCA's great contribution to microelectronics technology was the development of a process technology called complementary MOS – CMOS. Although you couldn't pack the transistors in as closely using CMOS as you could with other technologies, CMOS had the advantage of requiring very little power and made possible chips for battery-powered products like watches and calculators.

RCA made the first CMOS transistor in 1963 and introduced the first chips using CMOS in 1968. Later on, CMOS was re-engineered by

Hitachi of Japan to give comparable performance to other technologies without losing any of its low-power characteristics. That's why CMOS became the microelectronics industry's standard production process in the 1980s and 1990s.

Taiwan has reason to be grateful to RCA for bringing in the island's first chip technology. In 1976 Taiwan was becoming a significant manufacturer of watches and calculators and it wanted to become self-sufficient in the chips that powered them. Obviously the technology it needed was CMOS, and RCA, as the inventor of CMOS, was the obvious source.

Taiwan set up a laboratory at Hsinchu City, about an hour on the train from Taipei, to take the transfer of technology from RCA. That was in 1976 and the technology level was sufficient to make 4 kilobit DRAMs. Today that same laboratory has developed 16 megabit DRAMs and is working on 64 megabit chips. Surrounding the lab is most of Taiwan's semiconductor industry.

RCA was swallowed up as a microelectronics supplier when GE took over the whole corporation in the 1980s. Within a couple of years of the takeover GE had sold both its own microelectronics division and RCA's chip business to Harris Corporation.

INTEL

Founded in 1968 by two of the eight Fairchild co-founders Robert Noyce and Gordon Moore, plus Moore's Fairchild deputy Andrew Grove, Intel was the first company to sell SRAMs, and invented the DRAM, EPROM and the electrically erasable PROM (EEPROM). However, the invention for which it is most famous, and on which its fortunes almost exclusively rest in the 1990s, is the microprocessor – invented in 1971.

The scale of Intel's innovation in the chip industry can be judged from the fact that the products it invented account, collectively, for about half the value of the chip industry in the mid-1990s.

One by one, Intel stopped making the products it pioneered as competition made them too expensive to manufacture profitably. However, one product it did not drop – the microprocessor – and in that product area it acquired an 80% share of the microprocessors used in personal computers in the early 1990s.

The success was founded on IBM selecting Intel's microprocessor for its PC, launched in 1981. That PC became the industry norm and every manufacturer of personal computers around the

world wanted to make IBM-type computers, so they all had to buy
Intel microprocessors. Intel astutely exploited that position by
cutting out its second sources (other licensed manufacturers) of
microprocessors, giving itself a virtual monopoly. That was how, in
1992, it became the world's No. 1 chip company – basically on the
revenues from a single product type.

Intel did a number of other things to reinforce its lead in microprocessors, but one of them was particularly unexpected – between
1990 and 1993 it spent half a billion dollars on consumer advertising to establish its microprocessors as a 'brand' in the minds of PC
buyers.

No one in the chip industry thought this was possible – they
thought it was like trying to brand a car engine – consumers don't
care about the engine – but it worked, and in 1993 the Intel brand
name was assessed by *Financial World* magazine as the third most
valuable brand name in the world after Coca-Cola and Marlboro.

MOTOROLA

One of only a few companies that have survived as players from
the transistor-making days to the chip days is Motorola.

Some attribute the success of Motorola as a chip-maker to the
location of its chip business in Phoenix, Arizona.

Phoenix is just about as far away from Motorola's corporate
headquarters in Chicago as it could be. The reasoning is that since
corporate bureaucracy stifles chip companies, the only way to
survive as the chip-making division of a large equipment company
is to be as far as possible from the corporate bureaucracy.

Motorola got into the transistor business early on as a support to
its equipment businesses in consumer electronics, communications
and defense. However, the company soon came to the crossroads of
either expanding into the merchant market to defray costs through
outside sales, or giving up transistor development altogether.

The company took the former course, and in 1955 was the ninth
largest transistor-maker in the USA. By 1956 it had 8% of the
market and was growing rapidly. By 1957 Motorola employed
around 400 people in its transistor factory and three years later it
had 2000 people producing 75 000 transistors a day.

From 1955 on, the company has always been in the top ten in the
world. It has particular strengths in making chips to support the
company's equipment divisions, e.g. in chips for telecommunications,

cars and radio equipment, but its greatest product is its range of microprocessors.

These were adopted for the Apple computer range, which accounts for some 10% of the PC business. This was not as lucrative a design win as Intel's in the IBM PC because Apple never licensed their technology. Consequently, no clone industry for Apple-type computers ever developed.

As well as microprocessors for computers, Motorola has been the most successful company in making microprocessors for non-computer applications, in everything from washing machines to robots. It has done this by successfully adapting its microprocessors to ever increasing numbers of uses. In the early 1990s it joined with Apple and IBM in developing a new micro, called PowerPC, aimed at pulling back the 80% market share Intel had established in the microprocessors-for-computers market.

NEC

NEC was founded in 1899 as a joint venture between AT&T and a group of Japanese investors. The company retains close links with AT&T, collaborating on basic research for chip technology during the 1990s.

However, AT&T sold out its stake in 1925 to Colonel Sosthenes Bean, Chairman of ITT, and NEC did not become wholly Japanese-managed until 1932, when ITT passed over management responsibility to the Sumitomo Group. NEC was not wholly Japanese-owned until 1960, when ITT sold its stake to Sumitomo.

Maybe because of its origins, NEC has been the most cosmopolitan of the Japanese chip companies, being the first to set up chip factories abroad. It has also cooperated with government research programs at home and led the 1980s 'charge' of the Japanese to catch up with the USA in chip technology.

Late in 1983, NEC started selling the 256 kilobit DRAM – a full year before any American company was able to get the chip out on the market. Japan had overtaken the Americans in the most technically demanding product in the chip industry – an event that was something of a trauma for the US chip industry.

NEC became the world's No. 1 because it is a big, broad-range supplier of most commodity chips, especially memory chips such as DRAMs, SRAMs and EPROMs. In the mid 1990s, NEC stepped up its strategy of manufacturing abroad with an aggressive program to manufacture a greater proportion of its chips in foreign countries. A billion dollar factory for 64 megabit DRAMs was started in Scotland.

TOSHIBA

Toshiba shot to the fore in the early 1980s, when a sudden collapse in the 256K DRAM market frightened other Japanese chip companies into cutting back on investment in the following generation of DRAM – the 1 megabit DRAM.

Toshiba maintained its investment and enjoyed a notable success as a result – for over half a year it was almost the only company in the world that could manufacture one megabit DRAMs in volume and it ended up with the lion's share of the market.

The company has been notably good at CMOS manufacturing and has cooperated with many international companies on technology exchanges and R&D programs. Partners include Siemens, Motorola, SGS-Thomson and IBM. It makes a broad range of products and has been in the top five world companies since the mid-1980s.

Toshiba dates back to 1875 and was originally called Tanaka Seizousyo after the founder, Hisashige Tanaka, who set it up as Japan's first manufacturing plant for telegraphic equipment. The name later changed to Shibaura Engineering and then to Tokyo Shibaura Engineering following its merger in 1939 with Tokyo Electric Company. This name was shortened, in 1978, to Toshiba.

For over sixty years from 1909, General Electric of the USA held 25% of Toshiba's shares. It bought them in 1909 and still held over 9% of Toshiba's stock at the beginning of the 1980s. During the 1980s, GE sold off its remaining holding.

HITACHI

Outside the ranks of US companies, the greatest contribution to MOS technology has come from Hitachi of Japan. The development of the MOS transistor in 1962 was a key contribution. It came about through the work of an engineer called Minoru Ono. His work was recognized by RCA, which encouraged Ono and Hitachi, and finally MOS became the mainstream technology route for the entire microelectronics industry.

Another key Hitachi contribution was the re-engineering of CMOS to convert it into a high-performance technology. It was a major contribution in the 1970s which made possible such 1980s products as portable computers and camcorders. Hitachi is also the leading exponent of BiCMOS technology – the art of merging the high speed of bipolar transistors with the low power of CMOS transistors on the same chip.

The company dates back to 1910, when the electrical repair engineer at a mining company built three 5 hp electrical motors for resale. Namihei Odaira, as he was called, set up a company for selling motors.

The company broke into the world's top ten microelectronics manufacturers in the mid-1980s and has been in the top five throughout the early 1990s. Although it makes commodity memories, like DRAMs, SRAMs, EPROMs and flash, it specializes in low-power versions or high-speed versions (or both), which means it can sell its chips for more than the standard going rate. The company also has its own line in microprocessors, particularly single chip 32-bit RISC (reduced instruction set computing) microprocessors, for which cost/performance has been greatly improved.

Hitachi has been more profit-conscious than most of the Japanese chip companies were in the 1980s and early 1990s. Starting in the 1980s it began sharing much of its fundamental R&D with Texas Instruments to defray the cost burden.

SAMSUNG

Samsung, the leading South Korean chip house, was a spectacular product of South Korea's industrial expansion in the 1980s. Although the conglomerate had some microelectronics manufacturing before the 1980s, it was small-scale and technologically backward.

The South Korean government was concerned that its burgeoning electronics equipment industry, mainly consumer equipment, had to import an ever-increasing amount of chips and was becoming dependent for many of these on Japan – with which South Korea had poor relations based on memories of military occupation.

So, in 1982, the South Korean government announced its 'Semiconductor Industry Promotion Plan' – involving grants, loans, special privileges and access to government-financed R&D. Also, low-cost loans from the World Bank were available.

The four companies chosen for the push into semiconductors were to be the four main industrial conglomerates – 'chaebols' as they are called in South Korea – Samsung, Hyundai, Lucky Goldstar and Daewoo.

Of these, Samsung had the most success. It did something unprecedented in simultaneously building factories for three generations of DRAM – the 64 kilobit, 256 kilobit and 1 megabit – and, largely based on the sales of memory chips, made it into the world top ten companies in 1993 – a remarkable achievement.

FUJITSU

Fujitsu entered the microelectronics business principally as a support for its equipment manufacturing. Fujitsu's parent company, Fuji Electric, had come into being as a joint venture between Siemens and the Furukawa Electric Company in 1923.

In 1935 Fuji Electric set up another joint venture company with Siemens to make Siemens' telephone exchanges in Japan. The joint venture was called Fuji Tsushinki which was abbreviated, in 1967, to Fujitsu.

In 1952 the company was asked to make a machine for automatic stock transfers and billing for the Tokyo Stock Exchange – in effect a computer. The team developing the machine was led by Toshio Ikeda. Although the Tokyo exchange never used the machine, the project had a lasting effect in that Ikeda got hooked on computers.

He cleverly got the company president to agree to continue computer development while he was absorbed in one of his favourite activities – watching a ballet.

According to Taiyu Kobayashi – later to be the company president – in his book *Fortune Favours the Brave*, Ikeda was an oddball, often arriving for work in the evening, which caused a lot of complaints and a loss of income for Ikeda, as staff were at that time paid by the day.

Ikeda assembled a team of people described by Kobayashi as 'curious individuals' and 'as if a number of stray kittens had been brought home and deemed so cute they were all kept around as pets'.

After visiting Ikeda's operation in the late 1950s, Kobayashi referred to it as 'a bunch of crazy exuberant engineers without any overall organization' working on five different types of computer at once. But they turned Fujitsu into the largest computer company in Japan and, in the 1990s, it became No. 2 in the world. The tradition of engineering-led company strategy has persisted.

Fujitsu's microelectronics operations were designed to make chips solely for computers and, in the 1990s, it remained the largest supplier of customized chips in the world. But the demands of defraying the costs of R&D pushed it into making standard chips such as memories. It has been one of the top ten chip suppliers since the mid-1980s.

MITSUBISHI ELECTRIC

Mitsubishi Electric of Japan was a top ten player in the late 1980s and early 1990s. The company is a member of the massive Mitsubishi conglomerate, which has diversified from its original activity, shipping, into banking, car-making, electronics and many other activities.

The Mitsubishi group started off as a shipping line back in 1870 and did not get into the electronics business until 1921 when it set up Mitsubishi Electric to make electric fans.

Mitsubishi Electric has been an efficient manufacturer of commodity chips with its automated manufacturing system. In the mid-1990s, it started devoting resources to the development of flash memory technology.

IBM

Until the late 1980s, the largest producer of microelectronics in the world – as well as being the largest purchaser of microelectronics – was IBM.

Until 1992 IBM kept all its production for its own internal use. In 1992, the company's declining fortunes in the computer business provoked the management into trying to cover some of its research overheads by selling its chip products on the open market.

Trying to turn the top-heavy managerial culture of IBM into a fleet-of-foot, flexible market performer was one problem for the company, another was trying to convert its products from being IBM-compatible to being market-compatible.

The first year IBM showed up in the microelectronics rankings, 1993, it was rated the tenth largest chip company in the world. But only about a tenth of its production was sold to the outside world.

IBM has a great record in microelectronics as both a researcher and a procurer. Its laboratories have always operated at the leading edge of pure research and it has been a powerful patron of innovation elsewhere. Many chip engineers have persuaded management, or venture capitalists, to invest in their ideas after an endorsement from IBM, and the company has been an enlightened source of procurement of new technologies.

For instance, it gave Texas Instruments an early multi-year procurement contract for the silicon transistor – a key inducement for Texas to keep on investing in the development and production of the device. That was a scenario repeated in many companies and over many new technologies, and it represented a powerful stimulus to the US microelectronics industry.

MATSUSHITA

Matsushita is the largest electronics company in Japan and the second largest in the world. It began in 1918 when, as a 23-year-old inspector with the Osaka Light Company, Konosuke Matsushita pondered on the inadequacy of the usual single power plug in domestic homes.

It cost him $50 to develop a double-ended socket and, in 1918, he founded Matsushita Electric Industry Co. Ltd to sell it. On the company's fourteenth birthday he assembled his 162 employees and unveiled his corporate plan – for the next 250 years!

Like Philips, Matsushita's electronics fortunes were founded on consumer electronics and its microelectronics division is primarily seen as a support to that division. None the less Matsushita sells enough general purpose chips, including memory chips, to be a major player.

PHILIPS

Europe has only ever had one company in the top ten, Philips of Holland, and to an extent that has been due to its ownership of the Silicon Valley company Signetics. Philips was a regular top tenner until 1993, after it had cut out a lot of its low-margin commodity lines in the search for profit rather than revenues.

Signetics had been one of the early spin-offs from Fairchild. In 1961, when Fairchild was only four years old and the chip itself was only three years old, a group from Fairchild decided to set up the first company in the world dedicated to making only chips.

By 1970 Signetics had become the fourth largest chip company in the world after Texas Instruments, Motorola and Fairchild.

Philips acquired the company in 1975 and has owned it ever since. Although Signetics gave the Dutch company a valuable pipeline into Silicon Valley thinking, Philips has always maintained European chip-making in addition to Signetics' factories.

Naturally, Philips' biggest chip products are chips used in consumer electronics – developed to support its consumer equipment division, which is one of Philips' main activities. Every now and again it ventures into the memory market, but the fluctuating financial fortunes of its parent have prevented it from consistently pursuing the erratic memory market.

Philips was founded in 1890 by Gerard Philips to make light bulbs. Later on, Gerard's brother Anton joined the company adding sales flair to Gerard's technical skill. From light bulbs it was a short step to vacuum tubes and from then to transistors and chips. Unlike many European electronics companies Philips always maintained its expertise in microelectronics at the highest level.

Philips was the company to which the young Konosuke Matsushita turned when he wanted microelectronics technology, and the result was a joint venture that lasted 40 years until it was bought out by Matsushita Electric Industry Co. Ltd in 1993.

NATIONAL SEMICONDUCTOR

Two 'Fairchildren' – the name given to companies founded by defectors from Fairchild – that hit the big league were Advanced Micro Devices and National Semiconductor. National was not quite typical of other Fairchildren in that it was already in existence as an ailing Connecticut-based subsidiary of Sprague Electric when, in 1966, Peter Sprague asked Fairchild's operations manager, a manufacturing wizard called Charlie Sporck, to take over as president of National.

However, Sporck reinvented the company in the Silicon Valley image – physically relocating it in California and turning it into the world's leading maker of analog chips – the chips that deal in electrical simulations of real-world signals such as heat, sound, pressure and weight. National became a top ten player within a few years of Sporck taking over, and remained so for nearly 20 years until the late 1980s.

Sporck diversified into memory chips and logic chips, and tried to establish a microprocessor without success, but the core of the company was the analog business – still the world's largest in the mid-1990s. Sporck ran the company with an airy indifference to the balance sheet, which in many years showed amounts of red ink that would have been alarming to any other CEO.

When the American chip industry went through a crisis of confidence in the late-80s, brought on by Japanese success in the business, Sporck was the prime mover in getting Sematech set up – a government-funded centre in Texas for developing new generations of manufacturing equipment and production processes.

SGS-THOMSON

Europe's No. 2 company in the chip business, SGS-Thomson, confounds a chip industry belief that mergers and takeovers don't work. That belief is based on the feeling that it takes smart people to succeed in the business and that, while you can take over assets, you can't take over people.

SGS-Thomson is a hybrid of the leading chip companies of Italy and France. It was put together in 1988 by the respective governments of the two countries and the merger has succeeded for two reasons: massive government financial and political support and an ebullient Sicilian President, Pasquale Pistorio. He had learnt the business climbing the corporate ladder at Motorola.

SGS-Thomson makes a broad range of chips, from memories to microprocessors and standard logic, and in the early 1990s was knocking on the door of the top ten. The corporate ambition is to get 5% world market share – roughly the market share of a top ten player.

SANYO

Sanyo was, in the 1993 Dataquest rankings, Japan's seventh largest semiconductor manufacturer. It is also a leading player in solar cells – for converting light into electricity – and has a major presence in advanced flat panel display manufacturing, with which it has a joint venture with Motif of Oregon, USA – a company partly owned by Motorola.

It has a strong capability in making the chips to support these activities, and also makes chips for TVs and other consumer applications.

In 1994, Sanyo decided to switch its investment from making DRAMs to making flash memory chips of the most advanced types under a license deal with Silicon Storage Technology of California.

SHARP

In the early to mid-1990s Sharp was the world's largest producer of flat panel screens for computers and TVs. Sharp joined up with RCA in the 1980s and Intel in the 1990s to expand its microelectronics activities. It has made DRAMs, but its speciality is ROMs, in which it is a leading player, and in the mid-1990s with Intel it is making a major play in flash memories. Sharp is also close to Apple and a licensee of the UK-developed ARM microprocessor.

ADVANCED MICRO DEVICES (AMD)

Advanced Micro Devices was one of the traditional 'Fairchildren'. In 1968, the same year as Noyce, Moore and Grove left Fairchild to found Intel, Jerry Sanders III and a team of Fairchild staffers defected to found AMD.

AMD grew almost as fast as Intel for many years – which was not surprising because it positioned itself as 'the leading follower' of Intel – making EPROM memories, which Intel had invented, and microprocessors, which Intel licensed to it. However, it branched out on its own into chips for telecommunications uses – and it became the world's top supplier.

Although AMD grew as fast and as big as Intel until the mid-1980s, its strategy of following Intel then suffered a major blip. Starting with the 386 microprocessor, introduced in 1985, Intel said that from now on no one but Intel would make it.

Sanders' main complaint was that for 15 years he had helped Intel establish its microprocessors and was now to be denied the fruits of those efforts. Bitter lawsuits characterized the next decade. AMD brought out its own versions of the 386 and subsequent 486 but the lawsuits delayed almost every plan. Still, even to be late to such an exploding market, it was still a marvellous business to be in and AMD prospered in the mid-1990s.

AMD became the world's largest maker of EPROM memories when Intel stopped making them, and is the second largest maker of flash memory. The company has been in the top ten but in the 1990s was hovering in the teens.

SIEMENS

A company that invested in AMD early on and made a lot of money from its investment as AMD grew to be a major player was the great engineering company Siemens of Germany. Siemens has always been in the microelectronics business as a second tier player but, until the mid-1990s, looked on it more as an expensive but necessary overhead than as something from which to make profits.

In the 1970s Siemens spent some $200 million (a fifth contributed by the government) trying to keep up in microelectronics technology. It was not spent to great effect. In 1981, Siemens launched its 64K DRAM – some four years behind the technology leader.

In the 1980s, Siemens engaged in a series of collaborations to keep up with microelectronics technology: first with Philips in an R&D joint venture, called 'Megaproject', for 1 megabit technology; then with Philips and SGS-Thomson under the JESSI banner for 4 and 16 megabit technology; then with IBM for 64 megabit technology; and then with IBM and Toshiba for 256 megabit technology.

The company dates back to the Great Exhibition of 1851, when Werner Siemens and Johann Georg Halske were recognized for their work in developing the telegraph. Siemens was early into lighting and installed the world's first electric street lighting at Godalming in the UK in 1881. The company has never compromised its basis as an engineering-based firm – which explains its persistence with microelectronics in the face of endemic unprofitability.

SONY

Sony's microelectronics operations date back to the licensing of transistor technology at the Bell Symposium of 1952 and the subsequent re-engineering of the device to make it suitable for portable radios. The Nobel Prize-winning 'Esaki-diode', named after its developer Sony's Dr Esaki, was part of this early work.

Sony has seen its microelectronics activity as principally a support for its equipment businesses, but in the 1980s it sought to defray some of its overheads by selling chips. Naturally its skills are in chips used in consumer goods, especially analog chips, and it is particularly strong in high-performance SRAMs.

The company was started in 1945 by an ex-naval lieutenant, Akio Morita, and his friend Masaru Ibuka, who had developed an adaptor allowing medium wave radios to receive short wave broadcasts.

The product that brought the company international recognition was the first transistor radio, and its most famous success – the Walkman – cemented Sony's position as one of the world's top three consumer electronics companies.

ROHM

Founded as a resistor company (hence its name, R-Ohm, named after the unit of measurement of electrical resistance), Rohm is the leading manufacturer of custom analog chips and owns the Silicon Valley custom analog specialist company Exar Corporation.

Rohm is also focused on electrically programmable memory technology and owns one of the Silicon Valley companies that pioneered the technology – Exel Microelectronics (a spin-off from Seeq Technology, which in turn spun off from Intel).

Rohm is primarily focused on consumer electronics. It sells standard logic chips and has developed memory chip technology, but has stayed away from the volume memory merchant market.

GOLDSTAR AND HYUNDAI

After Samsung, the two biggest South Korean chip companies are Hyundai and Goldstar. In the 1980s, both invested heavily in chip factories to manufacture commodity memories – DRAMs, SRAMs and EPROMs – as the basis for quickly growing into significant-sized companies.

Goldstar later went to Hitachi as a partner for its design expertise and subsequently improved its performance, starting to climb up the league tables. By 1993 it had reached No. 22 in the world league table.

In 1994, Hyundai followed Goldstar's example, forming a cooperation with Fujitsu for design expertise, and started on an aggressive program to become a top five player by the mid-to-late 1990s. In 1993, it was the world's 25th largest semiconductor company.

The South Korean forte is in managing massive factories and running them at a low overhead cost. They have not shown an interest in innovating, but, once they have established themselves in the commodity areas, their strategy is to move into the higher margin, higher performance areas and look for profits as well as growth.

Living with the Chip

OKI

Only five years after Alexander Graham Bell patented the telephone, Kibataro Oki set up a company in 1881 to manufacture telephones in Japan. Oki Electric remains at heart a communications company to this day, but it has branched out into other areas of electronics, including chips.

Oki mainly makes chips for telecommunications, but is also involved in memory chips and other commodity chip products, which it sells in the merchant market.

MICRON

Perhaps the most individualistic company in the chip business is Micron Technology. First of all, it's unusual because it's based in Idaho, whereas the rest of the US chip industry is in California, Texas or Arizona; secondly, it's unusual because it's a medium-sized company making DRAMs – normally a product made only by the largest companies; thirdly, it's unusual because it is backed by an eccentric billionaire who made his fortune selling chip potatoes to McDonald's; and fourthly, it's unusual because of the way it got started.

Back in 1978, when the UK government had set up a state-backed chip company called Inmos, approaches were made by the Inmos management to the best team of memory designers in the world. They were working for Mostek, a company in Dallas, Texas, which had grown rich on the back of two products – the 4 kilobit DRAM and the 16 kilobit DRAM. So good were the designs that much of the chip industry dropped its own versions of these chips and adopted the Mostek design.

Led by Ward Parkinson, who had been a schoolboy science prodigy, the team left Mostek to work for Inmos, developing the next generation of DRAM – the 64 kilobit, which was to be Inmos' first product. After a bitterly fought lawsuit, in which the Texas judge asserted he surely couldn't tell a 64K DRAM from a donkey's hind leg, the team of designers abandoned Inmos and took a design contract from Mostek, having formed themselves into a separate company called Micron Technology.

A year or two later, the son of the potato chip king, who was a friend of Parkinson, got his father, J. R. Simplot, to back them with a microchip factory in Idaho. Ever the patriot – he flew a 60 foot

US flag at his mountain-top home – J. R. decided that he was going to do something about backing American technology.

Micron had its up and downs, but its backer did not desert them and they pulled through to become a medium-sized chip firm, reaching half a billion dollar annual sales in the mid-1990s. Ward's brother Joe was a lawyer who helped out in the Inmos/Mostek lawsuit and became chairman of the company. As is the way of these things, the lawyer stayed on long after the scientist, Ward, had departed.

LSI LOGIC

LSI Logic was a company set up in 1979 specifically to go for the customized chip market. It followed the example of American Microsystems Inc. (AMI), which was set up for a similar purpose in the 1960s. LSI prospered on the back of significant advances in the technology of chip design by computer, which made it cost-effective to customize logic chips for specific applications.

However LSI, like AMI, could not sustain a consistently profitable business based entirely on custom chips and looked around for standard chip products. Meanwhile, eating away at the viability of custom logic chips was an architectural advance that threw up a new breed of chip that could be churned out as a standard product by chip-makers and then programmed by users to their own particular purposes.

These new programmable logic chips became significant as a market force in the late 1980s/early 1990s and look like becoming a more important logic chip type during the decade of standardization between 1997 and 2007 projected by Makimoto's Wave (see Chapter 5).

TEMIC

A second major force in German microelectronics emerged in the 1990s. The massive Daimler Benz organization, which owned Deutsche Aerospace and AEG Telefunken, among many other subsidiaries, decided to rationalize all its chip-making activities in 1992. Putting them all together in one company called Temic resulted in a chip business with total sales of a billion dollars.

It was an unusual procedure for the chip business, where it is generally assumed that entrepreneurial, flexible, locally managed, autonomous groups are more effective than large organizations.

HARRIS

Harris Corporation of Florida, the big US defense and aerospace contractor, has always operated a chip division. It has a heavy emphasis on special chips for defense and space needs and in the early 1990s absorbed the chip-making divisions of both General Electric of the USA and RCA. It is a medium-sized player focused on applications rather than the glory of large market share.

SANKEN ELECTRIC

Specializes in chips that go into products using high power, such as home appliances, cars, computer peripherals and TVs. It is strong in making analog chips. In 1993, Sanken was the 30th largest semiconductor company in the world, according to the Dataquest rankings.

HEWLETT-PACKARD (H-P) AND DIGITAL EQUIPMENT CORPORATION (DEC)

H-P and DEC are two major computer companies with sizable chip operations, but whereas H-P sells to the merchant chip market, DEC made everything for its own use until the mid-1990s, when it started to reconsider its position. Both companies use their chip factories to make their own proprietary microprocessors, which give their equipment its distinctive features.

However, in the 1990s both companies tried to get outside companies to take up their microprocessors in a bid to establish themselves as the owners of a standard industry architecture – just as IBM's adoption of Intel's microprocessor established it as the standard for personal computers.

The microprocessor DEC wants to establish is called 'Alpha' – blazingly fast and the performance king; the microprocessor H-P is trying to establish is called 'PA' for 'Precision Architecture'.

Most of H-P's sales on the merchant chip market come from sales of chips for optoelectronic purposes – i.e. chips that are used in equipment that involves fiber optics. These are cables containing thousands of fibers made of glass down which pulses of light can be sent to do the same job in communications that electrons down copper wires conventionally perform.

ANALOG DEVICES, LINEAR TECHNOLOGY AND MAXIM

National Semiconductor has been the No. 1 analog chip company for many years and, although every major chip company has an analog capability, there are three analog specialist companies selling chips on the merchant market: Analog Devices, which has been in the business since the 1970s, and two 1980s start-ups – Linear Technology, which was founded by a team from National, and Maxim Integrated Products.

Analog chip design is regarded in the microelectronics industry as a black art practised by an arcane priesthood of often wildly eccentric characters. What the analog people are basically engaged in is simulating electrically the physical phenomena associated with real-world signals. For instance, the vibrations in the air caused by speech are simulated by varying an electrical signal to replicate them.

Having simulated the signal electrically, it can then be converted into digital form, and vice versa. So a large part of analog companies' work is in making chips which convert analog signals into digital signals (A–D) or vice versa (D–A).

GEC-PLESSEY SEMICONDUCTORS

Analog is one strength of GEC-Plessey Semiconductors, which is the only UK chip-maker of significant size. It derives from the amalgamated chip operations of three UK chip-makers – Ferranti, GEC-Marconi and Plessey.

Plessey made the first-ever model of a silicon chip demonstrated at the 1957 International Symposium on Components. It was a non-working model, but similar to Kilby's 1958 working chip. Ferranti invented the semi-custom chip – where a chip is three-quarters made as a standard part and then has the last quarter customized to a particular purpose. Marconi developed world-class radiation-resistant processes for chip-making.

Strong in customized chips, GPS has mostly avoided making standard programmable chips. Having been in the microelectronics business from its earliest days, GPS has a strong position in intellectual property.

CYPRESS

Three Silicon Valley success stories that started in the mid-1980s were Cypress Semiconductor, Integrated Device Technology (IDT) and Cirrus Logic. Cypress was started by a former AMD engineer, the outspoken Dr 'T. J.' Rodgers. Cypress made a big initial impact in the late 1980s, making very fast SRAMs and programmable logic chips.

Rodgers quickly established himself as a national figure in the USA, promoting the entrepreneur as the engine of growth in the economy and deriding national industry support efforts such as the chip industry consortium Sematech, which he dismissed as a 'turkey'.

In the early 1990s, Cypress went through some tough times, having devoted a lot of resources to microprocessors – a trap for many companies – and Rodgers even sought technical advice from the derided Sematech. However, Cypress got back on course to be a profitable player in the mid-1990s, selling half a billion dollars worth of chips a year.

INTEGRATED DEVICE TECHNOLOGY (IDT)

One of the most successful Silicon Valley model start-ups in the 1980s was formed by a spin-off team from Hewlett-Packard called Integrated Device Technology. It made its fortunes in fast SRAM and a specialized breed of very fast microprocessors called 'bit-slice'.

IDT probably devoted too much of its resources to that thorny old trap for aspiring chip-makers – general-purpose microprocessors. It backed one of the new breed of 1980s 'RISC' microprocessors – the MIPS architecture. IDT has grown into a medium-sized player turning over several hundreds of millions of dollars.

CIRRUS

Cirrus Logic, founded in 1984, also grew to half a billion dollars in its first decade. Its brief is simple: to make every kind of chip needed in PCs except memory chips and microprocessors.

It proved a great formula as the PC gradually expanded its capabilities. Starting with chips to connect up the hard disk to the PC, it then made chips to connect the screen to the microprocessor and the keyboard to the microprocessor. It then went for chips adding graphics, then video, then wired and wireless communications, then linking up CD-ROMs to PCs. As the clone-makers took over the PC market, Cirrus Logic boomed by supplying them with the chips to keep their products up to date.

ALTERA, XILINX, LATTICE, ACTEL AND MONOLITHIC MEMORIES

Five companies pioneered the development of programmable logic chips, each of which took a different technological approach: Altera, Xilinx, Actel, Lattice and Monolithic Memories.

Altera was one of the 'Fairchildren' – its founding team spinning off from Fairchild shortly after it was bought by the French oil-field services group Schlumberger. It developed programmable logic chips based on EPROM technology. So they became programmed by the user, but only reprogrammable if put under ultraviolet light and erased.

Xilinx, a spin-off from the microprocessor pioneering company Zilog, used SRAM technology, which meant that its chips could be readily reprogrammed. However, it also meant that they used a lot of transistors (it takes six to make an SRAM cell whereas it only takes one to make an EPROM cell). However, as the market developed, Xilinx took the lead.

Lattice developed programmable logic using EEPROM technology i.e. EPROM that could be erased by an electrical signal, which is a lot more convenient than using ultraviolet light. Using EEPROM, the chips could be wiped and reprogrammed without disturbing them or the equipment into which they are designed and it could be done from afar – even if the equipment was in space.

Actel's route to making programmable logic chips used a web of electrical circuits in which at every connection – where one circuit line met another – there was a capacitor.

The capacitor acted as an insulator, stopping electrical flow, but it could be made a conductor by putting a large enough charge down the line, so melting the capacitor and turning it into a conductive line. By deciding which capacitors to melt and which to leave in place, users could program the chips to their own purposes. But, once programmed, the chip could not be reprogrammed.

Actel calls its technology 'anti-fuse'. That's because there was a 1970s technology for producing programmable chips, invented by a firm called Monolithic Memories, which had a fuse at every interconnection on the matrix of circuit lines. In that technology every connection was a good working connection until an electric charge was sent down the line to blow the fuse and so turn that connection into an insulator. Known as 'fuse' technology, it prompted Actel, using the reverse technique, to call its technology 'anti-fuse'.

Like many chip companies, Monolithic Memories had one good idea. It stayed with its specialty, not branching out into other chip areas, had its heyday in the 1970s and was taken over by AMD in the 1980s.

SHINDENGEN, RICOH, TOKO AND FUJI ELECTRIC

Shindengen is a leading player in making microelectronics for cars and motors; Ricoh makes microelectronics principally for its range of copiers, faxes, printers and telephones; Toko is concerned with microelectronics in so far as it relates to handling large electric currents; and Fuji Electric is mainly interested as the leading supplier of vending machines, as a major player in solar cells and as a growing player in factory automation.

SEIKO-EPSON

The chip-making arm of the leading printer-makers, Seiko-Epson's main concern is to produce chips to support its printer products. However, it has become a notable silicon foundry source – offering leading technology manufacturing processes to companies that do not have chip factories of their own.

CYRIX

A small company in the early 1990s, Cyrix has a lot of potential as a designer of versions of the most profitable line of chips in the world in the late 1980s and mid-1990s – Intel's 386, 486 and Pentium microprocessors.

Cyrix has, none the less, found appropriate growth difficult to achieve because of legal actions from Intel and from its first manufacturing partner, Texas Instruments.

Not being able to afford, in the first half of the 1990s, a chip factory of its own, Cyrix found it difficult to get enough chips to sell from either its first manufacturing partner – Texas Instruments, or its second manufacturing partner – SGS-Thomson.

Frustratingly for Cyrix, it had the hottest product in the industry – chips that sold for hundreds of dollars each in a market measured in tens of millions of units – but it couldn't get enough of them made. In 1994 it signed up a third manufacturing partner – IBM.

The problem for Cyrix was one that afflicts many microelectronics companies without a factory – manufacturing partners will only agree to manufacture their chips in return for the rights to make and sell them on their own account. Naturally, these manufacturing partners tended to become more interested in making chips for themselves than for Cyrix.

Though beset by lawsuits, Cyrix was still battling on in the mid-1990s as a quarter of a billion dollar company with the ever-present potential of becoming a huge one.

CHIPS AND TECHNOLOGIES

A company that had a major effect on both the chip industry and the computer industry – effectively ending IBM's control over PC design – was a company called Chips and Technologies.

It was all the brainchild of Gordon Campbell, formerly of Intel and Seeq Technology (where he was founder and president). After leaving Seeq, Campbell set up C&T, whose first product was a set of five chips costing $70. They did the same job as 63 chips in the then current IBM PC. Anyone wanting to make a clone of an IBM PC could do it using Campbell's chips and end up with a faster, smaller, cheaper PC than IBM's. The product took the market by storm.

However, Campbell over-reached himself in deciding to produce not only the chips surrounding the microprocessor in PCs but also the microprocessor itself. Too much of the company's resources was put into 'reverse engineering' the microprocessors. Then Intel sued again, taking up more resources, and competitors came in to commoditize the PC chip-set market, driving down prices and profitability. The company went through bad times in the early 1990s.

ITRI AND ERSO

By the mid-1990s Taiwan had become the fifth largest regional player in the chip business after America, Japan, Europe and South Korea, with half a dozen companies of significant size and growing fast.

Taiwan evolved a very effective model for stimulating its chip industry. Based on a laboratory in Hsinchu City owned by the state body Industrial Research Institute (ITRI), it developed the CMOS technology it had bought from RCA in 1976.

ITRI's electronics arm, Electronics Research and Services Organization (ERSO), from time to time set up prototype production processes to test out the technology developed in the laboratory at ITRI on real-life commercial products.

Once these products looked good and the team producing them had mastered the technology and the production process, ERSO then spun off the production line, the team and the products into fully fledged commercial companies.

UMC, TSMC AND WINBOND

By the early 1990s, three companies had got started that way: United Microelectronics Corp. (UMC), Taiwan Semiconductor Manufacturing Co. (TSMC) and Winbond Electronics Corp.

TSMC is an unusual company in that it is a semiconductor company without products of its own but which acts as a manufacturing base for the many pure design houses in Taiwan. It has become the world's most successful silicon 'foundry' – somewhere anyone can go to get their designs turned into silicon.

Taiwan's two biggest chip companies, UMC and TSMC, reached half billion dollar annual sales by the mid-1990s, when they built billion dollar chip factories.

The third company to get started under the ITRI/ERSO formula, Winbond, is on the same growth path as UMC and TSMC, but has been more concerned than them in developing innovative technology. In pursuit of new ideas, Winbond has formed alliances with, and bought into, a number of Silicon Valley start-up companies.

In 1992, ERSO started the whole cycle up again with a new production line to test the half-micron process developed in ITRI's lab. It has been an effective model for getting technology from the lab into industry, for nurturing successful commercial companies and so growing a powerful local semiconductor industry.

The model is successful because it takes the heavy start-up burdens of unlimited expense and unforeseeable technical problems off the shoulders of the start-up team. Once they have a working technology and competitive products and they are able to survive in the outside world, then they are set free to fly.

Taiwan has taken a different route from the Japanese and South Korean route into the industry – they have concentrated on logic chips rather than memory chips. Logic is a much smaller market than memory, so Taiwan's growth has been slower, but logic profit margins are better, the logic market is less erratic than the memory market and logic appeals to the Chinese strength in design.

MOSEL-VITELIC, HUALON AND MACRONIX

Besides UMC, TSMC and Winbond, the other major Taiwan chip companies are Mosel-Vitelic, Hualon and Macronix. Mosel-Vitelic is the result of a merger between the island's SRAM specialists, Mosel, and its DRAM specialists, Vitelic. The company has substantial backing from the Japanese companies Sumitomo, Oki, Kyocera and Sony.

Hualon Microelectronics is unusual in that it is independent of ITRI and ERSO – in fact, aggressively so, having, according to ERSO, recruited ERSO's technologists to acquire technology but declining to pay ERSO any license fees.

Hualon was the result of a stock market flyer taken by the textile company Hualon when the Taipei stock market was booming in the late 1980s. The stock exchange was particularly keen on semiconductor companies, which were seen as having unlimited upside potential. Hualon recovered its cost of starting up its microelectronics subsidiary within a year of founding it by selling around a third of the shares.

Macronix is another product of the late 1980s Taipei stock exchange boom – raising $160 million in equity and loans in 1989. Its engineers are mostly returnees from the USA – Taiwanese who have studied and worked in the USA. Their main aim is to be a major player in flash memory, but they started making many different kinds of chip while waiting for the flash market to develop.

ATMEL

When Gordon Campbell left Intel to set up Seeq he took with him the brothers Perlegos. In 1985, the same year Campbell left to start Chips and Technologies, they set up a company called Atmel to pursue electrically erasable PROM (EEPROM) technology. Diversifying into programmable logic chips, they had a significant growing company by the mid-1990s.

ZILOG

One of the most extraordinary companies in the chip industry is Zilog, which made the best-selling microprocessor in history but no profits. The company was started by Federico Faggin, who had put Ted Hoff's original concept of the microprocessor into silicon at Intel. It was sold as the '4004' – the world's first microprocessor

– a 4 bit chip, which meant it could handle four binary digits (a 1 or a 0) at one time.

Faggin then teamed up with Hal Feeney to make the first 8 bit microprocessor, the 8008, and then led the team that developed the 8080. However, after the 8080 development, Faggin left Intel to set up his own company. He called it Zilog and it developed the world's best selling microprocessor – the Z80.

The Z80 is still being designed into new products, like Amstrad's first 'personal digital assistant', which was brought out in 1993. So prevalent is the Z80 in Asia that it is commonly assumed there to have been invented by Japan.

However, having invented his chip, Faggin let everyone make it, and Zilog never made a cent in profits for its first ten years. And when the company moved from the 8 bit Z80 to the 16 bit Z8000 – a much better designed part than Intel's 16 bit microprocessor, which was adopted for the first IBM PC – the company failed to support it effectively in the market with the necessary tools to help designers use it.

Zilog was bought by the oil giant Exxon with a view to diversification, but Exxon could not make money out of it. In the 1980s, a management buyout brought Zilog into profitability for the first time in its history – but it never repeated its innovatory triumph of the Z80.

Cracking the microprocessor market was regarded as a Holy Grail in the 1970s but no one ever managed to dislodge Intel and Motorola from their dominant positions in providing microprocessors for computers – the largest and most lucrative market for microprocessors.

INMOS

The most bizarre attempt to crack this perennial hard nut and attain the chip industry's Holy Grail of establishing a new microprocessor came in 1978 when the Labour government in the UK used a state agency responsible for promoting industry, called the National Enterprise Board, to back an attempt to make a new kind of microprocessor.

The plan was to set up a company that would get established in memory chips and then develop a microprocessor. The microprocessor was to be aimed at removing the computer industry from the tyranny of the ubiquitous Von Neumann architecture – i.e. one microprocessor per computer processing one chunk of information at a time – and allow multiple microprocessors per computer processing multi-chunks of information simultaneously, i.e. parallel processing of data instead of serial processing. The intention was to improve computer performance radically.

The company was called Inmos, and it had a notable success, not with microprocessors, but with a new concept for SRAMs, which provided its only substantial revenue stream. The microprocessor, when it came, puzzled the industry. It had its own programming language, which alienated many, and, although it found many uses outside computers and was used in some specialized supercomputers, it was never used in ordinary computers sold in shops.

Perhaps that was because Inmos never recognized the old adage that people buy applications – not sexy technology – and no one wrote useful, affordable software applications packages to run on the Inmos microprocessor, called the 'Transputer'.

Like Zilog before it, Inmos never made any profits. The UK government recovered its investment by selling it to Thorn EMI, but Thorn lost hundreds of millions of dollars owning Inmos before passing it on to SGS-Thomson.

Before getting the UK government to fund Inmos, the company's founders had tried to sell the idea to the government of South Korea, which was also looking, in the late 1970s, at ways to get into the microelectronics industry.

ARIZONA MICROCHIP

With similar technologies and products to its parents is a company called Arizona Microchip, which was spun off from General Instrument when GI decided to exit the chip business. It had good EEPROM technology and used this to move into programmable microcontrollers.

NIPPON STEEL, KOBE STEEL, KAWASAKI STEEL AND NKK

The Japanese steel companies, Nippon Steel, NKK, Kawasaki Steel and Kobe Steel were all looking enviously at the semiconductor industry in the late 1980s. Whereas the steel business was flat or declining, the semiconductor business had average annual 20% growth rates – something undreamt of by the steel industry for many a year.

Japanese steel companies saw the semiconductor business as being not unlike their own – process driven, capital intensive, highly expensive – and decided to diversify into it. Although they could build factories however, it was more difficult to come up with designs, and during the first half of the 1990s they tended to make products for other people.

XICOR

A company that has stayed successfully exploiting a niche is Xicor – one of the early spin-offs from Intel, hence its name: (e)x-i(for Intel) + cor(p). (It takes an engineering mind-set to think of these names.) Xicor did things the nice way – asking Intel for a technology license and paying for it.

The technology Xicor paid for was an earlier version of the electrically erasable PROM technology which was later taken by Gordon Campbell (without payment) to set up Seeq Technology.

The Xicor technology was very different from Seeq's and it was not suitable for very high-capacity memories. However, it was very useful to people who wanted to give their products some element of reprogrammability – like a door chime, answerphone or aeroplane black box, or for providing speech in products.

MIETEC AND AUSTRIA MIKRO SYSTEMS (AMS)

Two typically European companies are Mietec and Austria Mikro Systems. They are typically European because they have a penchant for difficult and exotic chip production processes – like handling very high power, or mixing analog and digital, or combining bipolar and CMOS. They are also typically European in that they actually dislike high-volume products – their whole structure is set up to make a good margin on unusual processes in small volumes.

Mietec is owned by Alcatel, the giant French company, which is the world's biggest supplier of telecommunications equipment. Austria Mikro Systems (AMS), grandly installed in Premstatten Castle, derives from the US custom chip company American Microsystems Inc. (AMI), which had been set up in 1966 as the first custom chip company, and was sold to Gould in 1982.

AMS remains a custom chip company and has therefore stayed small. It is one of the basics of the industry that growth comes from making standard programmable chips in volume – a truism that derives from the high fixed overhead nature of the industry's underlying economics.

MIKRON, ANGSTREM, ELECTRONICA AND INTEGRAL

The four top Russian chip companies are, in the mid-1990s, moving their production from making standard memories and microprocessors mainly for domestic and military customers to making commercial chips for the world market.

Mikron and Angstrem, both of Zelenograd near Moscow, have been aggressive in overseas markets with watch, games, calculator and ASIC chips, and are moving into higher value-added areas, such as telecommunications chips for mobile communications.

Electronica of Voronezh is a specialist in power semiconductors and is strong in chips used in cars. Integral of Minsk, in Belarus, has been one of the most successful exporters of chips from the former Soviet Union and is particularly regarded for its logic chips. It also makes watch chips and memory chips.

These are a few of the great, near-great, late-great and wannabe-great chip companies. There are at least a couple of hundred more around the world, and the number always grows. Their fortunes wax and wane with the shifts of technologies and markets and the vagaries of investors.

Many companies live on the verge of triumph or disaster – and seem to thrive on the risk and the excitement: 'Some people pay to go to a bull fight or watch *Terminator* films', muses Grove of Intel, 'we get paid for living it'. Some of the more dramatic triumphs and disasters are in the next chapter.

7

Triumphs and disasters

Hubris, said the Greeks, precedes disaster and in the chip business there is always plenty of hubris. How could there not be hubris in 20-year-old chip designers who see their products sold in quantities worth tens of millions of dollars, or in 30-year-old company presidents who run hundred million dollar companies that they founded, or in 40-year-old executives signing off for billion dollar factories?

The archetypal story of Silicon Valley hubris is the fate of the 40-year-old Dennis Barnhart, president of Eagle Computer. On the day his company was to go public, putting a value on his personal shareholding of $9 million, he and a friend went out for a spin in his Ferrari, lost control on a bend and Barnhart was killed.

But the greatest hubris displayed in Silicon Valley was not Californian hubris or even American hubris. It was French hubris. Schlumberger was a company, specializing in the oil-field services area, that had diversified into many areas and had always diversified successfully.

In 1979 Schlumberger bought what it regarded as the jewel in Silicon Valley's crown – Fairchild Semiconductor – the company that had invented the industry's process technology and that shared the patent on the chip.

Schlumberger saw Fairchild as the star turn of an industry that was expected to grow ten times in the 1980s (it didn't), and Schlumberger had a cash mountain of over $2 billion to invest in that growth.

It looked a great scenario, but it wasn't. The jewel was tarnished by the time Schlumberger bought it – Noyce, Moore and Grove had left a decade earlier to found Intel and, the same year (1968), Sherman Fairchild had responded by persuading the entire top management team of Motorola's semiconductor division to take over the management of Fairchild.

The price for leaving the Arizona desert for California's orchards astonished the entire industry. In addition to his $120 000 salary, Motorola's president, C. Lester Hogan received a $5.4 million interest-free loan to exercise an option on 90 000 shares at $60 each, and a further allotment of 10 000 shares at $10 each. The paper profit on the shares, before the year was out, was $2.5 million. It was a substantial package for 1968.

However, the Motorola management style had not transferred successfully to California. Hogan and his top team – dubbed 'Hogan's Heroes' – had taken Motorola's semiconductor division from $5 million to $230 million, but they did not do much for the Californians.

In his first few months Hogan made the costly mistake of sidelining the formidable Jerry Sanders III, who walked away with some of the company's top talent and founded Advanced Micro Devices, which was soon to become an industry leader.

Losses in 1970 were followed by more losses in 1971. The company made the wrong bet on the future direction of memory chip technology. Hogan bet on a new technology called the charge coupled device (CCD), when the technology that the market actually went for was MOS (metal oxide semiconductor) technology.

Fairchild's struggle to master MOS – without which it could not address the fastest growing areas of the chip market – paralysed the company's prospects. Between 1968 and 1975 there were five different managers of the MOS operation. None of them solved the problems.

The problems did not get any better after one of the 'Heroes', Wilf Corrigan, went over Hogan's head to the president of the foundation that controlled a large chunk of Fairchild stock and threatened to resign if he didn't get Hogan's job. Corrigan got the presidency, but he did not have the answers to Fairchild's problems.

He diversified into the digital watch business, where prices were soon to collapse, and into computers. The two moves lost around $70 million. And even in Fairchild's core business, semiconductors, in the six years 1973–9 the company slipped from the No. 2 position in the industry to No. 6. In 1979, in the all-important MOS area, it was 14th.

In the 1970s, the market for MOS-based products had grown from 6% of the world chip industry's output to around 35%. Although Fairchild had versions of other companies' MOS products, which was a low-profit way of competing in MOS, it could not master the design and manufacturing of its own proprietary MOS products, which was the only way to attract high-margin MOS revenues. One engineer walked out claiming: 'They can't build MOS worth a damn'.

Between 1975 and 1979, Fairchild's chip sales grew at little more than half the industry's annual growth rate. Fairchild no longer set the industry's pace – instead it was Intel, Motorola and Mostek, with their MOS-based memories and microprocessors. Fairchild was, according to an analyst at Montgomery Securities 'the lousiest of the majors'.

However, all these cracks were not immediately discernible to the outside world, because Fairchild still had considerable revenues from its royalty income, and that was pure profit which went straight to boost the bottom line.

As the inventor of the planar process – the fundamental production process that everyone in the industry had to use – Fairchild could demand royalties from every company in the industry, and the fast-growing industry was constantly spawning new companies. Consequently, the balance sheet looked strong.

And in a good industry year, like 1979, when Fairchild grew by nearly 30% to clock up revenues of nearly half a billion dollars, the company could still look impressive. That was the year Schlumberger bought Fairchild for $400 million. But beneath the stately superstructure of the good ship Fairchild, the barnacles were encrusted and growing and threatening to stall her progress.

Whether Schlumberger knew of or understood the problems besetting its acquisition is uncertain; what is certain is that the company believed that it had two great assets to bring to the company – cash and management expertise.

The smack of firm Schlumberger management expertise came swiftly and as a severe shock to the laid-back Californians. 'Jean Riboud [Schlumberger's Chairman] called us all together and said "You guys have been having an easy time. Now you're going to work your butts off". Then he cancelled all the bonuses', said one appalled Fairchild senior manager.

Laid-back Californians in the chip industry, used to being highly valued and knowing they could get jobs anywhere in the industry, weren't going to take talk like that. Five general managers walked out.

Out of the 20 most senior managers in the company, seven left in the first twelve months of Schlumberger's ownership. Some went to plum jobs elsewhere – George Wells to be president of Intersil, Rodney Smith to be president of Altera, Gil Amelio to be president of Rockwell (later to be president of National Semiconductor), Jim Bowen to be group vice president at Eaton, and Brian Sear to be president of Genrad's semiconductor division.

In place of the Fairchild management style, Schlumberger slowly and painfully built up a new structure based on rigid hierarchies, headed by an ex-West Pointer, Thomas C. Roberts. It was all so painstaking and happened so agonizingly slowly that competitors dubbed the company 'Slumberchild'. In an industry where products became obsolete in three years, taking endless time reorganizing management structures was regarded as eccentric if not foolish.

Schlumberger scrapped the Fairchild tradition of paying over the odds for the best people and abolished the 'key man' bonus scheme in favor of the Schlumberger philosophy of loyalty to the company. In a free-wheeling industry where job-hopping was endemic, loyalty was an anachronism.

So in one of the two areas of strength Schlumberger was to bring to Fairchild – management expertise – the contribution was almost wholly negative. The other area of strength was money. With a $2 billion cash mountain in 1979, Schlumberger had the financial muscle to reinvigorate Fairchild – so it thought.

In its first full year of ownership of Fairchild, 1980, Schlumberger put over $100 million into capital spending and R&D. That trend continued. In 1982 the figure for R&D was $105 million and for capital spending was $188 million – nearly $300 million in all.

The money was poured into five upgraded or new American factories and into a new chip plant in Germany. It also poured into new product development. In spending so much money, Schlumberger recognized the importance of two key chip industry requirements: up-to-date process technology and a continual stream of new products.

New products attract the highest margins, and one way of getting a better performing product than the competition was to have better processes using finer line widths and smaller transistors to increase speed and performance.

That much about the industry Schlumberger clearly grasped. But it failed to recognize the importance of smart people. Having a better performing product than the competition is no good if the industry simply doesn't want that kind of product any more. Only smart people, betting correctly on the future course of the industry, could make sure that the new product budget was being spent on new products that the market wanted.

And only smart people, recognizing the enormous difficulties of getting complicated new products and processes to work, were capable of knowing whose suggestions to listen to, whose hunches to back and where to devote the effort and resources. And most of the smart people had gone.

'Success in the semiconductor business takes people, process and money', according to Ben Rosen, co-founder of the venture capital firm Sevin-Rosen, which backed Compaq, Lotus and Cypress, among a host of successful high-tech company start-ups. Process and money Schlumberger could deliver; people it could not.

Recognizing that problem in 1985, Schlumberger went, as Sherman Fairchild had gone almost 20 years before, for a total management transplant. It recruited a team of Texas Instruments' top managers. There were so many of them that the new industry tag for Fairchild was 'TI-West'.

The TI-ers could do little to stem the decline of Fairchild. By 1987 over $1 billion had been poured into product development, research, upgrades to factories and new factories. Added to the purchase price, the acquisition of Fairchild had cost Schlumberger some $1.5 billion. In 1987, a new chairman of Schlumberger decided that enough was enough and put Fairchild on the market. He sold it for $122 million. The venture into microelectronics had cost Schlumberger some $1.4 billion.

Fairchild co-founder Gordon Moore observed, 'Schlumberger was a victim of hubris. They'd been successful with everything they'd touched but they got Fairchild at a time of deterioration – I'd far rather start a company from scratch than try and fix a sick one'.

Starting a chip company from scratch is not trauma-free, as Gordon Campbell will testify. The first time he tried it with Seeq Technology he got a writ in the middle of the night from his employer – Intel. Then, four years later, after building a profitable

$40 million a year company from scratch, and paying their original investment back to his venture capitalist investors, they sacked him.

The second time he tried it was also pretty traumatic. Campbell decided his next move after Seeq would be to make chips to provide the electronic innards of personal computers. Having spent most of the $2 million he'd received for his Seeq shares on leasing giant mainframe computers to design the chips, and having had the idea turned down by so many venture capitalists that he was becoming convinced he'd been blackballed by the financial community, Campbell eventually got $1.5 million backing from a Japanese group of investors.

That set the scene for the most meteoric success of any chip company in the history of the industry. 'To me it was bound to fail', says the man who introduced Campbell's products to Britain, Nigel Williams, 'it was a company no one had ever heard of and they were asking computer makers to rely on them totally – no one was going to buy their product.'

Williams was wrong. Apricot Computers of the UK bought Campbell's chips, recognizing that the world was moving towards a single standard for desktop computers and that standard was the IBM PC.

Apricot did not have an IBM-compatible computer and to design and make one would take so much time that Apricot feared it would miss the market. Campbell's chips offered a short cut. Around the world a good many computer companies went through the same reasoning loop as Apricot and committed themselves to Campbell's chips.

The chips cost around $70, there were five of them, and they did the same job as 63 chips in IBM's own version of the PC. The only other IBM proprietary ingredient needed was the Basic Input/Output System (BIOS), which connected up the software operating system to the hardware machine. The BIOS had been reverse engineered (copied) – perfectly legally – and incorporated in a chip (a ROM), which could be bought from a company called Phoenix Technologies of Boston. Campbell's salesmen helpfully provided the address.

They also provided a complete board layout showing PC-makers exactly how to design their computer board so as to provide total compatibility with IBM PCs. All that the computer-maker had to do was go out and buy some memory chips from the Japanese or South Koreans, a microprocessor from Intel and a license for the MS-DOS operating system from Microsoft.

Campbell contracted out the manufacturing of the chips to factories in Japan, America and Europe. By using a dozen different
factories Campbell could keep control of manufacturing costs and
supply without having to commit to the expensive overhead of chip
manufacturing.

The first chip-set replicated the IBM-AT computer. Whereas
IBM's version ran at 8 MHz, a PC made with Campbell's chip-set
ran at 10 MHz, used a third of the power of the IBM and required a
third of the chips of an IBM, so halving the size of the computer
board.

The formula was a staggering success. The only capital
Campbell's company – called Chips and Technologies – had ever
raised totalled $3 million. The company started in 1985 and paid
back its investors by April 1986. In October 1986 it went public,
raising $10 million, which valued the company at $70 million.

At the end of June 1986, the company had sold $12.7 million
worth of chip-sets for a gross profit of $7.7 million. By June 1987
it closed its financial year with sales of $80 million and a gross
profit of $25 million. Revenues-per-employee were $650 000.

The value of Campbell's 3 million shares in the company soared
to $70 million. Over in the UK a jubilant Nigel Williams had,
against his better judgment, sold $7 million worth of Chips' chips
in a year.

Visiting Chips' Silicon Valley offices in those days was an eye-
opener. The place was full of chip technologists figuring out how
to second-guess IBM in the next generation of computers. So smart
were Campbell's engineers that they could bring out chip-sets that
anticipated IBM's products – allowing their customers to bring out
IBM-compatible PCs ahead of IBM itself. Moreover, they
performed better.

However, a visitor being taken round Chips' HQ by Campbell
might just as well have been taken round by the janitor – so
absorbed in the game were the engineers that the president seemed
like an irrelevance.

While Campbell's clever Californian engineers figured out the
design problems, on the other side of the Pacific PC clone-makers
were solving the problems of the low-cost manufacturing of high-
tech products.

It was a classic combination of the economic strengths of the
two regions – America with its entrepreneurial system throwing up
free-wheeling, independent design talent of the highest caliber and

the Far East wedded to economic growth at all costs based on dedi-cated, competitive, high-quality, low-cost manufacturing.

Neither could have succeeded without the other, but together they produced a whole new industry of IBM-cloners, and from that moment on IBM took an ever-decreasing share of the market for PCs based on the standard it had set.

By 1988 the PC market had so exploded that around 15 million of them were being sold every year. By 1990 it was topping 20 million. The value of the market soared from the billions of dollars a year to the tens of billions. The prospects for Campbell's chip-sets seemed unlimited.

In its third year of trading, 1987, Campbell's chip sales shot up again to $141 million, for a gross profit of $35 million. There was, however, a threat looming. IBM, fed up at seeing its PCs cloned by all and sundry without getting so much as a license fee or a royalty, decided to bring out a line of uncloneable machines. They were called the PS/2.

By using a good deal of patent-protected proprietary-design hardware, IBM hoped to make the PS/2 range uncloneable. Campbell got round that by taking the published functionality of the chips used in the IBM machines and getting his engineers to design chips that performed the same functions but were totally different internally in their design. This was held by the US law courts to be a perfectly legal approach.

IBM announced its 'uncloneable' PCs in April 1987, and by August Chips had announced it had a chip-set cloning the bottom of the range Model 30. Moreover Chips' chips ran 25% faster than IBM's. By February 1988, Chips announced it had similarly cloned the Model 50, with a 30% performance premium, and by April it had a clone chip-set for the top of the range Model 80, which ran at 20 MHz compared with the 16 MHz of IBM's chips.

The following year, 1988, Chips reported revenues of $217 million with a gross profit of $51 million. Next year, 1989, Chips' sales hit the quarter billion mark. No company in Silicon Valley's history had reached the magic $250 million mark in five years. Even Intel had taken ten.

This being a tale of a triumph, we'll leave the Chips and Technologies story here. Suffice it to say that what Chips could do to IBM – essentially commoditize its PC – other people could do to Chips.

Venture capitalists, suddenly seeing they had been wrong to universally snub the chance of making millions with Campbell, put their money into funding a dozen new companies like his.

The rash of new competition commoditized the PC chip-set business – bombing prices and over-supplying the market. It was the oldest story in the book for the microelectronics industry.

Still Chips was a triumph. It went against the received wisdom of the time; it took on the biggest names in the industry – IBM and Intel; it invented an entirely new segment of the chip industry – no one had made PC chip-sets before; and it was the fastest growing company in the chip industry's history. That should be enough to be a triumph.

Now for a disaster. Back in 1969, a new chip company was started up in Carrollton, a suburb of Dallas, Texas. Its name, as with many chip start-ups, was meant to say something about the company. For instance, Intel stood for 'Integrated Electronics', Siliconix stood for 'Silicon Integrated Circuits'. This company's name was Mostek, and MOS technology was the industry's hottest new obsession.

The date is significant. It was the year after Intel had been founded to pursue MOS technology and the same year Hogan was recruited from Motorola to galvanize Fairchild's attempts to master MOS technology. So Mostek was making a pretty public statement from the start about its intentions in the chip business.

Mostek was founded by a team from Texas Instruments, led by L. J. Sevin, later to found the venture capital firm Sevin-Rosen. With Sevin were Vin Prothro and Berry Cash. They positioned the company for the highest growth, highest risk markets in the chip business – memories and microprocessors.

Mostek took part in the 1K, 4K, 16K and 64K DRAM generations. It was one of the first companies to bring out a 4K DRAM and was one of the first two (the other being Intel) to come out with a 16K DRAM. It also made top-selling microprocessors, such as Zilog's Z80, Intel's 8086 and Motorola's 68000.

To all intents and purposes it was a leader both in the technology and in the market, seemingly effortlessly mastering the MOS technology over which so many rivals struggled. By the end of the decade Mostek was a top ten player.

That was enough to attract the attention of large corporations which envied the growth rates of the semiconductor industry. Moreover it was becoming fashionable in some high-tech corporations at the end of the 1970s to believe that it would be necessary to have top-class in-house microelectronics technology if they were to remain competitive in electronics-based equipment businesses.

For instance, the electronics giant Gould made bids for both Fairchild and Mostek in 1979, losing (fortunately for Gould as it turned out) in both instances to bids from other large corporations also convinced by the argument that they needed to own their own source of microelectronics expertise.

The large corporation that took over Mostek was United Technologies Corporation (UTC) – famous for its subsidiaries Pratt and Whitney (aeroplane engines), Sikorsky (helicopters) and Otis (elevators).

UTC paid $380 million for Mostek in 1979 and the following year Mostek's sales were over $370 million. However, almost immediately the market turned down. In 1981, sales slumped to $296 million and Mostek was suddenly losing millions of dollars. The reason was the old microelectronics problem of too many producers oversupplying the market, in this case the DRAM market. In the consequent price slump of 1981, Mostek ended losing $100 million. It was the first of many tabs that UTC was to pick up.

Other expenses that UTC seemed, initially, happy to pick up were for two very large new chip factories in Colorado and Texas, an $80 million new assembly factory in Dublin, Ireland, a joint venture with the then financially troubled AEG-Telefunken of Germany, new assembly plants in both Texas and Malaysia and a United Technologies' Microelectronics Center in Colorado which aimed to use Mostek's chip skills to work with UTC's equipment skills to develop computer-designed chips that would add competitiveness to UTC's electronic equipment products. That, anyway, was the theory.

The spending list looked a bit as if UTC regarded Mostek as an intriguing new toy. Certainly the financial controls it exercised over its new subsidiary seemed pretty lax. In his book, *No Excuses Management*, T. J. Rodgers, President of Cypress Semiconductor, recalls interviewing a Mostek chip designer for a job at Cypress in 1983. 'He showed me a chip photograph of his project. He had written in the scribe line in 10 micron aluminum letters "This RAM has been made without the knowledge of management".'

'The bootleg project syndrome at Mostek was one of the sicknesses that eventually did the company in', commented Rodgers, who regarded the Mostek engineers as 'a bunch of arrogant prima donnas.'

Notwithstanding the prima donnas, UTC kept putting in the money to keep Mostek competitive, and the company established a lead in the 64K DRAM market. In 1983, revenues recovered to $315 million and between 60 and 65% of that came from the 64 kilobit DRAM.

The reliance on the 64K DRAM had its benefits in 1984 when the industry grew faster in one year than it ever had before. That year Mostek recorded its best-ever sales of $450 million, but its product portfolio was unbalanced. That took a dreadful toll when the disaster year of 1985 struck.

It was a year when the worldwide microelectronics industry lost $6 billion as over-capacity in the memory market drove prices down to ridiculous levels and when all the major US DRAM producers, except Texas, pulled out. For Mostek, with its heavy reliance on DRAMs, the loss was worse than for anyone.

For UTC it was totally traumatic. The company had already invested $1 billion since taking over Mostek in new factories, in capital equipment, in R&D and in covering losses.

In return for that billion Mostek would, hoped UTC, bring out a 256K DRAM in 1983 – ahead of all its competitors – and so take a technological lead in DRAMs and grab the lion's share of the early, most profitable, revenues in the DRAM market.

However, technical problems delayed and delayed the launch of the 256K and then, in December 1983, NEC of Japan beat them to the punch by announcing a 256K up and running in production.

So when, in the first six months of 1985, Mostek lost $215 million on sales that had halved, and was clocking up losses of over $1 million a day that looked unstoppable, it was not surprising that UTC gave up believing that Mostek could solve its own problems.

In a final act of desperation UTC paid out $2 million to hire the No. 2 semiconductor man at Motorola, Jim Fiebiger, to sort things out. In November that year, Fiebiger concluded a deal with Thomson of France, selling Mostek to them for $71 million. After the $380 million purchase prices and the subsequent billion dollar investments in the company, Mostek had cost UTC some $1.3 billion.

When the Thomson men went into Carrollton to see what they had bought, they found 80 unused wafer steppers – the most expensive production tool used in chip manufacturing costing then about $500 000 each. Some had not even been taken out of the crates used to transport them from the manufacturers to Mostek.

The story of Mostek demonstrates the importance of positioning and timing in the microelectronics business. By getting MOS technology under control early on Mostek could produce high-value-added products with little competition and at a good margin. UTC was unlucky in that it bought Mostek just before the big Japanese assault on the DRAM market. And Mostek had become dependent on DRAMs for over half its revenues.

But if engineers could make products without the knowledge of management and if unwanted half million dollar machines could be ordered and left idle, then UTC must take some blame. The kids appeared to have taken over the school.

After a disaster, it's salutary to look at a triumph, and of all the triumphs in the chip industry, none has been quite so glittering as Intel's. Like Mostek, it got started in the late 1960s and, like Mostek, it set out to make MOS technology a viable commercial technology. Unlike Mostek it was innovative outside the memory field, diversified its products, stayed independent and was still being run by its founders 25 years after start-up.

The orderly climb of its revenues make it look an effortless progress – $2,672 in 1968; $66 million in 73; $400 million in 78; $1.1 billion in 83; $2.8 billion in 88; $8.7 billion in 93 – but Intel had its moments of anguish, being bailed out by IBM in 1982 at a time of great financial trouble and going through a very bad patch in 1985 when the employee count was cut by a third, pay was cut by 10% and compulsory days off without pay were introduced.

Being led by engineers with the self-confidence of a great track record the company has been prepared to bet its future over and again on a particular vision of the industry's future. For instance, the R&D spend was maintained in the disaster year of 1985, and in the recession year of 1986 the R&D spend and capital investment combined accounted for a third of the company's revenues.

Similarly the company has had the self-confidence not to become wedded to products as an article of faith and to get out of them before commoditization sets in. In 1982 it pulled out of SRAMs – the company's first product, which it had made back in 1969 – and in 1985 it pulled out of DRAMs.

'It was an emotional decision', said co-founder Andy Grove. 'We had been the first to introduce the product and build the business.

Even as we were losing market share hand over fist, we clung to the idea that we'd come back.' DRAM was taking a third of Intel's R&D dollars and contributing only 5% of its revenues. Pulling out freed resources to go into microprocessors. It was, said Grove, 'one of the toughest and the best decisions we ever made'.

Again in 1992, Intel pulled out of EPROMs – a product it had invented – even though it was the world's No. 1 supplier of EPROMs at the time. The reason was a decision to devote its factory capacity and its development dollars to flash memories – EPROMs that could be erased electrically rather than by ultraviolet light. Intel was also No. 1 in flash: it saw better margins on flash than EPROM, it saw fewer competitors and it saw that flash, not EPROM, represented the future. That's logical, but few companies give up a business in which they are world No. 1 just because it's the logical thing to do.

Intel showed that the high-tech business was not immune from the forces that affect all businesses. It became adept at using politics, the law courts, the media and the advertising agencies to maintain and expand its position.

However, it did not lose its belief in its engineering and market judgment or the self-confidence to bet the company on it. Three factories, each costing around a billion dollars, for manufacturing the 1994 Pentium microprocessor were built before the product was launched. Few company presidents will take that sort of gamble – if they want to sleep at night.

Besides Intel, two other companies of late 1960s vintage went on to become billion dollar companies by their fifteenth birthdays – National Semiconductor and Advanced Micro Devices (AMD). Both were led by remarkable, though very different, men: Charlie Sporck of National and Jerry Sanders III of AMD.

Sanders's company was split between Texas and California and he kept a black Rolls-Royce in Texas and a white Rolls-Royce in California. Asked why, he said, 'So I know where I am'.

In contrast, Sporck drove a pick-up truck to work and was the epitome of the tall, lean American Westerner doing his own thing in his own way. After the industry disaster of 1985 he led the US microelectronics industry in a bid to obtain political support for a combined effort in R&D to remain competitive in the face of seemingly overwhelmingly strong competition from Japan.

Behind the majors are a number of successful companies that got started in the 1980s and were at or around the half billion mark by the mid-1990s: Micron Technology, Cirrus Logic, Integrated Device Technology and Cypress Semiconductor.

Behind them were up-and-comers such as the programmable logic pioneers Xilinx, Atmel and Altera, with sales in the $100–200 million range, but growing fast with the growth of their particular market.

The great microelectronics companies of Japan had their triumphs in the 1980s when they seemed unstoppable. Six of them moved into the top ten, and they internationalized their operations, moving factories into America and Europe.

However, the recession of the early 1990s, which hit the Japanese chip market harder than anywhere else, caused the Japanese a difficult time, as their domestic market collapsed. By then NEC, Toshiba, Hitachi, Fujitsu and Mitsubishi each had microelectronics businesses worth between $3 billion and $6 billion a year.

Their concern, in the mid-1990s, is how to move into the higher margin areas of the chip business as the commodity areas fall first to the South Koreans, followed, it is expected, by mainland China.

Perhaps the most extraordinary triumph – if it can be called that – was the success of Texas Instruments in turning its intellectual property in the industry into its most profitable business area. In the late 1980s and early 1990s Texas started picking up royalty revenues worth hundreds of millions of dollars a year.

Having been in the microelectronics business since the early days of transistor manufacturing, Texas had a host of fundamental patents relating to the industry. The normal industry practice was to trade off patents between companies, swapping the rights to one chip design or process technology for another.

However, in the late 1980s a new president at Texas, Jerry Junkins, took a new tack, instructing his lawyers to go out and maximize revenues from the company's intellectual property portfolio.

A notable source of revenue came after the Japanese Patent Office recognized Jack Kilby's patent – some 30 years after it was filed in the USA – even though, in the mid-1990s, Fujitsu was still fighting in the courts against the validity of the patent.

From royalties on the Kilby and other patents, Texas extracted a vast stream of revenues which increased through the early 1990s and which appear to be interruptible only by time, when the rights expire.

Texas' success in tapping so lucratively into its intellectual property has led people to suggest that the company may be pioneering a model for the high-tech industry. It is suggested that companies could exist simply by selling ideas – not requiring factories or even any tangible products – simply selling pure intellect.

In a sense – in software companies, which sell nothing but intellect – we have that already. But that is applied intellect. What is beginning to look increasingly viable as an economic model for the industry is a company that sells ideas for products and lives off its licenses and royalties.

An example is Rambus Corp. of the USA, which developed and licenses a new method of connecting microprocessors to DRAMs so as to provide a faster interchange of information between them. Another example is Pilkington Microelectronics of the UK, which develops and licenses a technology for making programmable logic chips. A third example is Advanced RISC Machines (ARM) of the UK, which licenses a microprocessor design. None of them has manufacturing operations or marketing organizations – they simply develop ideas and license them.

Whether or not such companies are the model for future microelectronics companies is debatable. Certainly the success of Chips and Technologies shows that manufacturing is not a necessary ingredient to success to microelectronics. On the other hand, a leading capability in the industry's process technology will always go a long way towards ensuring a company's success.

The triumphs and disasters of the microelectronics industry give a flavor of why some companies succeed and why some others don't in this most unpredictable of industries. If people, process technology and money are indeed the three keys to the industry then it will remain unpredictable.

For while money always seems to be available for microelectronics from a variety of sources – governments, large companies and venture capitalists – and while process technology is almost ubiquitous in the modern world, people will always be an unpredictable element.

Time and again the industry throws up the lesson that motivated, committed clever people succeed, but cleverness by itself, motivation by itself, or commitment by itself is never enough because the microelectronics business is an industry in which change is endemic but unpredictable.

The microelectronics race very often does not go to the big or the strong, but to those with the sensitive antennae to pick up on change and the flexibility to act to exploit it.

The intrinsic human fondness for things-as-they-are, and the natural human resistance to change, have proved fatal to success in a business in which products become obsolete in a couple of years, and where new ideas can change the fundamentals of the entire industry.

If the triumphs and disasters of the microelectronics industry tell us a lesson it is this: that open-mindedness to every new idea and an open ear for every piece of industry gossip are pre-requisites for survival. It was summed up by Andy Grove: 'In the semiconductor business, only the paranoid survive'.

8

Living with the chip-makers

Investors in microelectronics companies have made fortunes and lost their shirts; employees of microelectronics companies have been made millionaires and laid off without a penny; customers of microelectronics companies have seen their products make millions and bomb hopelessly.

Consumers, buying microelectronics products in High Street shops, have been delighted by games machines and given new earning capabilities by home computers, and have then watched in horror as goods they paid heavily for have dropped in price to a fraction of what they paid.

Like it or not, most of us increasingly have to deal with microelectronics companies – whether as customers, investors, employees or purchasers of the products into which microelectronics are put.

For investors, California is Mecca. Ever since William Shockley set up his laboratory among the fruit trees of the Santa Clara Valley, microelectronics engineers have seen California as the route to riches. And not only American engineers. The boardrooms of the entrepreneurial start-up companies of the Valley are the most cosmopolitan assemblies of people to be found anywhere in the world.

Following, a hundred years later, in the footsteps of the Forty-niners, the protagonists of the Silicon Rush have come from all over the world – particularly India, China, Taiwan, South Korea and Europe – partly to find an environment where an engineer can do what an engineer must do, but mostly to get rich.

The Valley has spawned literally thousands of millionaires. The earliest, and most dramatic, example of almost instant riches came when the eight men who founded Fairchild Semiconductor in 1957 had their shareholdings bought out by the parent company for $250 000 each in 1959.

One of the Fairchild eight, Gordon Moore, then invested that $250 000 in the new company he co-founded in 1968 – Intel. Twenty five years later that investment was worth $1.5 billion. In 1994, Moore's 23.2 million Intel shares represented 5.6% of the company's equity.

The founders of companies that get a successful product to market are almost certain to become millionaires and, with the invariable Valley rule that all employees hold stock, it is inevitable that everyone in a successful company gets some capital reward and the early employees get pretty rich.

With these kinds of glittering rewards as examples, it is no wonder that the world's ambitious chip-makers flock to California. Although many other areas of the world have tried to replicate the conditions that make the Valley such an entrepreneurial success, no other country ever has.

It probably has something to do with the Californian genius for making life as simple as possible. For the chip-maker, everything is on hand from the financiers of San Francisco to lawyers, estate agents, accountants and PR firms that understand the microelectronics business.

But, more importantly, all the essential support services – test houses, materials analysis houses, equipment makers – are all on the doorstep to provide expertise in the multifarious disciplines required for the highly complex business of chip-making.

Essentially, it all comes down to the fact that it is simply easier to start up a chip-making company in California than anywhere else – moreover, the world's largest chip market is on the doorstep.

That is why the microelectronics community generally believes that if you go to a financier – whether in London, Frankfurt, Tokyo or New York – with a business plan for starting a chip company anywhere outside California, you will normally be turned down.

But if the embryo operation has in its postal address the magic words Santa Clara, Palo Alto, Mountain View, Milpitas or any other of the districts of Silicon Valley, then at least you stand a chance.

Herman Hauser, who co-founded the UK computer company Acorn in the 1980s, and then founded a company to make personal digital assistants in the 1990s, tells how the same proposition was turned down by venture capitalists when headquartered in England but accepted when the HQ's address was changed to Mountain View.

So the rule for investors in the 1990s is, as it was in the 1950s, 1960s, 1970s and 1980s, put your money into a Californian company. Investors at the early stages of a company's lifetime have a basic rule – look for a good team. More important than a good technology sell to the mind of the early investor is the belief that the founders of the company have a track record in working together.

Once the company is up and running then the investor is looking for execution – how well the company does what it set out to do – whether it meets its design schedules, its product development deadlines and its manufacturing timetables.

At this stage everyone in the company, whether it be investors, founders or employees, will be working towards one goal – the public offer. This is when the company gets a worth placed on it by selling off a chunk of its shares to the general public and getting a listing on the stock market.

With a listing on the stock market, anyone with shares in the company can sell them for ready cash whenever they like. Successful flotations bestow millionairedom on the founders and substantial riches on the key employees. The investors would probably be looking to see ten times their original investment back after the usual three to four years from founding a company to its public offer.

The success of public offers is more about timing than the performance of the company. Companies that have done a super job at bringing out excellent products and selling them profitably have had a disappointing response from the public share offer; other companies have raised enormous amounts of money on little more than promises.

The reason is simply that microelectronics companies' shares are sometimes fashionable and sometimes not. Hit the market when chip stocks are hot and your public offering can make you a fortune – when LSI Logic went public it raised over $160 million before it had made a penny in profit. However, if you hit the market when chip stocks are in low esteem, your shares will be valued low, even if your company has done a first-class job.

It is often said that it doesn't matter what you put in the offer document – the document that describes the company to potential investors – so long as you put in everything and it is accurate. The determinant of whether or not the stock will be bought is not going to be the shrewd appraisal of a knowledgable public – few people are qualified to judge such companies – but simply fashion.

After the company has gone public and has a listing on the stock market, future investors will be looking for something else – technological breakthroughs. News of a breakthrough in microelectronics technology at a company, or news of a hot new product, will send the stock soaring – even if commercial exploitation is months, or even years, away.

The reason why investors are looking for breakthroughs is because they are investing for growth. No one invests in chip companies for dividends – very few chip companies pay dividends anyway. It was over 20 years from its foundation before Intel paid a dividend. So if you're investing for growth you want to see companies coming out with blockbuster new high-margin products that will boost revenues and profits and grow the company.

That is why investors watch very carefully the amount of money that microelectronics companies spend on research and development. R&D expenditure measures the effort going into future new products and an early indicator that a management is foreseeing trouble ahead is the fact that it has cut back on its R&D spend.

Less R&D means fewer new products, and since it is the new products that produce the high margins in the microelectronics business, the clear indication to potential investors is that a cutback in R&D will lead to lower margins in the future. Lower margins means less money available to spend on R&D in the future, and so on and so on.

An early indication that a large chip company foresees trouble is if it delays or fails to build factories to make its highest technology products. It will normally be necessary to start building a factory for a new generation of product about eighteen months before the product is due on the market. If potential investors see a company deciding not to go ahead with such factory building it is a fair assumption that the company has switched into 'safety-first' mode – a sign that growth will slow down.

Similarly, if a company starts to pull out of the most demanding technical areas (e.g. DRAMs) it is often a sign that the management has decided that it is time to retrench and retract. Unless, that is, the management has found newer – higher growth – areas where the investment would be better spent.

Another key sign of danger is if a company, especially a small company, decides to develop general-purpose microprocessors for computers. Microprocessors are to the microelectronics industry what black holes are to cosmology. Companies have lost untold millions of dollars investing in microprocessors for computers, always forgetting the lesson learned by IBM in computers many years ago – that customers want compatibility, not change.

It has been proved, over and again, that a microprocessor that is technologically superior to anything else on the market has not got a hope of success if it cannot run the great mass of the existing applications packages that the owners of computers use.

And not only that, in order to make it worthwhile to the computer-maker to switch microprocessors, the new microprocessor would have to run the existing applications packages faster than existing microprocessors if they are to make the switch, and/or have a significant price advantage.

So the lesson to all potential investors in microelectronics companies is to beware of the company that is going into the mainstream microprocessor-for-computers business. Up to the mid-1990s, no one except Intel and Motorola has succeeded in making money from it.

Another indicator that a chip company faces trouble ahead is if it starts talking about switching its focus from standard products to customized products. Customized products usually require less than leading edge technology and customers often pay up-front for the work, making it a safe option for microelectronics companies.

The most important indicator of trouble that investors should watch for is if key employees start leaving. The key employees know they can get jobs anywhere in the industry and they commonly do not want to stay with companies where the urge to pioneer the technology and grow the company has been abandoned for the sake of caution.

One or all of these indicators signal that the management is looking for safety-first measures rather than to grow themselves out of trouble with new products.

What the potential investor should be looking for is a company with a plausible vision. Vision is a very important thing in the microelectronics industry. It involves developing products ahead of when the market requires them – but not too far ahead.

For instance, it is a fair bet that the two tools on which the consumer electronics industry will be based for the next decade are the personal computer and the TV. These will gradually expand and merge their capabilities. So chips that add the characteristics of one to the other are going to be required.

Chips that will connect the computer to the phone or to broadcast transmissions, and chips that connect the TV to compact discs and magnetic disks are the sort of areas in which companies in the 1990s are finding exploitable market niches.

Another 1990s area where there are opportunities for plausible visions is in the transmission of digital data, where the rapidly falling costs and rapidly rising speed of transmission (in terms of bits per second) are providing microelectronics companies with the challenge of providing the chips that perform such functions as the compression and decompression of information.

So the key for anyone buying shares in a microelectronics company is the plausible vision: have they got a credible strategy for future growth and are they making the necessary investments to achieve it? If they don't plan to grow, then there's absolutely no reason to buy their shares.

For customers of microelectronics companies, i.e. people who make electronic equipment and need chips to provide the electronic innards, the prime consideration should be much the same as for the investor, i.e. 'Is this company in it for the long haul?'.

If the company is not making the necessary investments for future technological capability and new products, then, whatever it says, it is not in the game for the long term. In the mid-1990s, the cost of investment in new generations of products is so huge that many of the largest companies are linking up with other large companies to share it.

So the measure of a large company's commitment in the 1990s is probably whether it has a credible partnership with another large company (or companies) to share the cost of developing the industry's basic process technology.

To go back to the analogy of the Forty-niners of the Gold Rush, whereas the early pioneers of the microelectronics industry found their nuggets of gold on the Earth's surface, waiting to be panned, by the mid-1990s 'You have to excavate two miles underground', says Hitachi's Jim Duckworth.

And excavating two miles underground is an expensive business. The amount of money spent to reduce line widths on chips from the 0.35 millionths of a meter (a mid-1990s process) to 0.25 millionths of a meter (a late 1990s process), to take the industry from the 64 million transistor chip to the 256 million transistor chip, is in the region of a billion dollars.

So the key to discovering the commitment of a large company today is to find out how, if at all, it is proposing to fund its process development, and the key to discovering the viability of a small company is how it is going to get access to the process technology of the large companies.

For the small companies this is a perpetual dilemma. They need leading-edge process technology to be competitive but they cannot afford it. Fortunately they tend to have something that large companies don't usually have – good ideas. So in many cases they will trade off a good idea to get access to process technology.

However, if the small innovative company gives away too much – for instance if it gives away the right to manufacture and sell its products – it creates a market rival who can probably afford to sell at prices lower than the small company can afford.

So the small company has a difficult path to follow. A sensible trade is often that the small company will allow the big company to use its idea internally, but not have the right to sell it on the market.

So investors have a number of questions to ask before they put their money into a chip company. How credible is the management? Has it got a credible strategy for growth? Has it got access to the latest technology? Has it got control over its products, i.e. are they protected by intellectual property law or by some over means? Has it got control over its prices?

Some of these questions should also be asked by potential customers of microelectronics companies. You might ask why it is important to choose suppliers carefully – certainly it never used to be.

In the original days of the microelectronics business suppliers routinely ignored quality and customers' delivery deadlines, and sometimes did not even bother to fulfill customers' orders at all if they found them inconvenient.

Furthermore, consignments of devices were sent out that the microelectronics manufacturers knew contained a large proportion of defective parts. Prices were whatever manufacturers could get for their parts – high in the good times, low in times of over-capacity.

The equipment industry responded by having no compunction in returning orders if changing circumstances meant it no longer had a use for the parts. In those days a written order for chips was not worth the paper it was printed on – both parties felt free to renege on its terms.

That made possible a practise that severely destabilized the microelectronics industry. Customers would double order or triple order their requirements from different suppliers, knowing that they could cancel orders without penalty if they became surplus to requirements. That way a customer of the microelectronics industry could survive periods of shortage.

This anarchistic state of affairs contributed greatly to the microelectronics industry's booms and busts, as manufacturers, elated by massive ordering, built new capacity only to see the orders evaporate into thin air as they turned out be double or triple orders made only to provide security in the event of rivals' possible defaults. The subsequent cancelation of orders left microelectronics companies with expensive factories to pay for and no demand.

The nakedly confrontational relationship between microelectronics manufacturers and electronics equipment manufacturers persisted into the late 1980s. By then, however, service issues had become quite as important as technology and price issues in the customer/supplier relationship.

Customers started to expect totally reliable quality – parts that did not even need to be tested by the customer on arrival. They also started to demand daily deliveries of the parts to their production lines so the devices could be put directly into products without having to go into stock. Furthermore, customers demanded, and got, absolute adherence to the specifications they required.

In return, the microelectronics companies got supply contracts that could be relied on and the two sides of the industry started to work as partners rather than antagonists.

One of the forces pushing equipment-makers towards closer relationships with their microelectronics suppliers was the trend throughout the late 1980s and 1990s towards products with shorter and shorter lifetimes.

This meant that the equipment-makers had to learn to trust the microelectronics companies to come up with chips in time to go into their own equipment products according to their planned production schedules. And the microelectronics companies had to learn to get their products out on time. To be late was to be useless in many cases.

So, towards the end of the 1980s a more trusting, collaborative relationship between the microelectronics companies and their customers became the norm, with each letting the other more and more into their future product plans. In that way the customers influenced the microelectronics companies in the types of chip they wanted, and the microelectronics companies helped define the equipment types that its customers would be able to make with the new chips.

By the mid-1990s the industry norm was close cooperation between chip-maker and chip-user at a number of different levels. For chip-makers, chip-users tended to fall into three types: innovators, early adopters and late adopters.

Innovators see their companies as being at the forefront of technology: they want the chip company's products as soon as they become available.

Some innovators offer to be what the industry calls a 'beta site' – a test site for testing the first versions of a microelectronics company's latest products. As such, innovators see themselves as being part of the creative process for developing new products.

At the other extreme are the late adopters. They are either risk-averse or work in an industry where the penalty for using a device that could have a bug in it, or where the supporting documentation is incomplete, is very severe. An example is the car industry, where recalls are hugely expensive while having the latest electronic technology is not so important that it's worth taking a risk for.

In between are the early adopters – wanting to be near the forefront of new product introduction, where the largest profit margins are made – but reluctant to take too much of a risk with unproven technology.

Since the early and late adopters will be watching the innovators to gauge their reactions to new products, microelectronics companies are usually very concerned to develop close relationships with the innovators.

Whether innovator, early adopter or late adopter, all customers should insist on one thing – a 'road map' showing the microelectronics company's intentions for future, better, more technologically advanced products. In a famous 1980 marketing campaign by Intel – Operation Crush – the company compiled a 100 page 'Futures Catalogue' with detailed specifications of all the company's future products.

That's the kind of commitment that makes customers feel happy. In the case of Operation Crush it succeeded in making 2000 of them happy enough to design Intel's microprocessors into their products during the year-long campaign. One of the 2000 was a design-in at IBM for its first PC.

The other main concern of a customer is price, and here the chip industry is notorious for its erratic pricing. Partly this is due to the nature of chip-making – when all the expense is at the beginning of the operation, and partly it is because of the erratic nature of the market.

The chip-making process starts off expensively because the first production of a new product is usually very unsuccessful. The chips – usually squares or rectangles – are made on round wafers with, usually, hundreds of chips to the wafer.

But whereas the cost of making the wafer stays pretty much the same, the number of working chips compared with the number of duds is usually pretty small at the beginning of the production cycle, meaning that the cost per chip is high.

As problems get flushed out, the proportion of working chips gradually grows, and so the cost per chip reduces. So the normal cost of producing a new product starts off high, then decreases.

However, most customers fall into the early and late adopter categories and usually need a bit of encouragement to move to a new product. So chip-makers, especially memory-chip-makers, tend to sell the early versions of new products at the sacrifice of profitability, keeping prices low in order to stimulate the market.

The problem for the chip-maker is exactly where to pitch this figure. If the price is too high, rivals can come in and take the market; if it's too low, then the chip-maker gets flooded with orders that can't be fulfilled, which annoys customers and encourages them to look elsewhere for an alternative product.

That's the chip-maker's problem. The chip customers' problems are security of supply and accurate predictions of future pricing. They will normally hope to see at least one alternative source of supply of any product they want to buy.

That has, in the past, encouraged chip companies to license their products to other chip companies to provide customers with more of a feeling of security.

However, chip companies have often found that their licensee is less concerned with maintaining reasonable profit margins than they are. With the originator of a product unable to control pricing levels – and not liking seeing the price bombed – there grew in the 1990s a trend towards the developers of new products keeping them to themselves.

This can be bad news for the customer because it results in higher prices. However, microelectronics suppliers will always have to make some concession to the learning curve process of continual price erosion if they are not to lose all sympathy with their customer bases, and the customers should always get some kind of commitment from their microelectronics supplier of the future pricing policy on a product.

This can't be totally predicted because no one can predict future demand or market-level prices, but customers should ask for, and get, an estimate of expected price erosion in normal market conditions.

Of course it could be said that, under normal market conditions in the microelectronics industry, there is either a shortage or over-capacity. It is pretty rare to see demand matching supply.

One reason for this is the huge expense of factories and the time it takes to build them. From the first building works it usually takes 18 months to get into first production, and from then on maybe a year to 18 months to get to full production. In that time anything can have happened to the market.

The old rule of thumb was to invest in new manufacturing capacity 'counter-cyclically', i.e. start building new factories at the worst moment in the market because – with the traditional four-year Silicon Cycle from market peak to market peak – two years after the worst moment in the cycle you would have a brand new factory to address the peak of the market.

That was how it used to be done, but the cycle itself appears to be changing, and the other factor that changes the game is the ever-lengthening payback time for new chip generations.

In the early 1980s it was a general rule that a microelectronics company would get back the money it had invested in building a new generation of memory chip in one and a half years. In the mid-1990s that had stretched out to four and a half years – with the 4 megabit DRAM only beginning to show a good profit on the investment in 1993 after it had first been moved into production in 1989/90.

Clearly, in that business environment, building any more manufacturing capacity than you absolutely have to is only going to make payback longer.

So it is likely that periodic shortages will continue to occur and, when that happens, the result is usually a flourishing 'gray market'. This happens because the respectable chip companies maintain their prices – not wishing to show customers they are exploiting them in times of shortage – but large quantities of chips get onto the market at sometimes quite exorbitant prices.

In these situations everyone blames everyone else. Chip companies blame their customers for selling their surplus chips to dealers to make a turn; equipment-makers blame chip companies for letting some of their production out onto the general market; and both of them blame distributors for trying to take advantage of the situation by selling some of their stocks at inflated prices.

However, for customers with production lines that could come to a halt without certain chips, there is sometimes no alternative to buying in the gray market. There are brokers, to whom they can turn, who will have contacts in the most thriving areas of the gray market – America, Hong Kong, Japan, South Korea and Taiwan.

The outward and visible manifestations of the gray market – such as advertisements for chips in the press – have contributed to spreading awareness among the general public of the value of chips.

It was not always so. In the mid-1980s, a van belonging to a major South Korean consumer electronics manufacturer, which was also a memory-chip-maker, was stolen in London. The hijacker drove it out of town before stopping to look in the back to check his haul.

He found a number of TVs and a box containing thousands of 256 kilobit DRAMs. He made off with the TVs but threw the box of DRAMs over the hedge. It was the wrong choice – the DRAMs were worth half a million dollars.

Never again. Since the early 1990s, the criminal fraternity has become only too aware of the financial value of chips. Chip factories and distributors around the world have been burgled for their stocks of chips and it is even said that organized crime uses chips – being small, valuable and easily tradable – as a highly portable means of transferring their illicit gains around the world.

By the 1990s, chips had become a readily tradeable commodity in the major cities of the developed world, just like precious metals or coffee beans, fulfilling the prediction of Jerry Sanders III, founder and president of Advanced Micro Devices, that: 'Chips will be the crude oil of the 90s'.

For most customers of the microelectronics companies, however, the normal rout to getting supplies of chips will be either direct from the chip companies or from their authorized distributors.

For these customers the questions to ask about a microelectronics supplier should be: are they easy to deal with? Are they accessible? Do they welcome me? Do they provide data easily? Are they
investing for the future?

So much for how customers should deal with microelectronics
companies. Another group of people with possibly an even greater
self-interest in choosing the right company is the potential
employee.

The classic choice for someone ambitious in the chip business is
between the thrill of the start-up and the drill of the established
company. The difference is between a results-oriented company
and an appearance-oriented company. In big companies you get on
by conforming to the company stereotype and playing the system;
in start-up companies you are measured solely by your effective-
ness.

The young company offers the chance to be fully involved with
every aspect of the operation, to get big responsibilities early on
and the opportunity to get rich via early stock options. The dream
of seeing your share options, acquired for a penny or two a share,
get valued at ten dollars in the initial public offering is the dream of
the employee in the start-up.

Even if you are not one of the very earliest employees, by join-
ing a fast-growing youngish company you have a better chance of
getting capital appreciation on your stock than owning stock
options in a large company where the share movements are much
more stable.

The downside of the start-up company is that it may go belly-up,
leaving you with nothing but a pile of unusable stock options. So
how do you choose a company that's going to succeed? The
answer is, in much the same way as you place a bet on a horse after
studying form. There are certain criteria for choosing a young
company to work for.

First of all there is the form of the founding team or the senior
management. Have they a good track record? Especially a track
record of working together?

Secondly, there's the backers. If the backers are well-known
high-tech venture capitalists – say Kleiner-Perkins or Sevin-Rosen
– that's a sign that the proposition has been run past the shrewdest
brains in the business.

Thirdly, there's the product. The key is: does it have a clear
competitive advantage in the market? And is that advantage main-

tainable either by keeping ahead technically, or by legal protection from intellectual property law or by some other means?

As with any microelectronics company, there must be a plausible strategy for profitable growth allowing for reinvestment in R&D – that's if you are ever going to get rich from cashing in your stock options. And getting rich is the main reason you joined a start-up in the first place.

However, not everyone wants to work 16 hour days or become a millionaire. For many the power that comes from achieving senior status in a large corporation is a sufficient reward; for others the security of a big company is important; and for most people there is little option to working for a big company because, outside the USA, there is not a lot of microelectronics start-up activity.

Choosing a big company to work for is also difficult because microelectronics companies tend to go through 'bullish' phases when they are investing and expanding, and 'bearish' phases when they are counting costs, trimming budgets and shedding staff.

The reason for that is that big chip companies are often divisions of large electronics equipment companies, and the boards of those companies decide what resources to allocate to the chip division. If microelectronics is out of fashion for some reason or other, a squeeze on resources can be imposed.

Clearly the smartest move for the potential employee is to join a company just at the beginning of the bullish period – that way there is plenty of scope for promotion as the company expands.

In the 1990s, a good many of the more dynamic companies are taking on the lessons of the American management gurus and are abandoning the old culture of hierarchical and directive management, with its many tiers of status, and adopting the model of a 'flat' management structure, with about five levels of hierarchy between the president and the shop-floor worker.

The thinking behind the 'flat' management structure is that the functions of the tiers of middle management can be better performed by computers. After all, the function of middle management is to act as a channel for information flowing upwards and downwards through the organization, and computers can perform this function – often more efficiently than people – making tiers of middle management unnecessary.

For instance, T. J. Rodgers, founder and president of a very successful 1980s start-up company, Cypress Semiconductor, personally monitors between 10 and 15 weekly goals for each of his 1500 employees through a software system he himself devised. He says it only takes him four hours a week to monitor the goals of all his employees and that it only takes each employee half an hour a week to review and update his or her weekly tasks.

In the same way, the best companies are devising computer systems that automatically inform all relevant departments whenever an event occurs that will require some action from them.

It is much the same sort of system that a modern supermarket implements, whereby, when a product is sold, the system automatically informs the stockroom, the transport department, the factory and all the other relevant departments of the organization of what needs to be done to replace it.

With the layers of middle management replaced by computers, the resulting 'flat' management structure means that the operational decisions are pushed down to those people who also have to implement them in practice. The result is that decisions get taken quickly – without levels of approvals to go through – and they get taken by the people who know all the practical problems involved in implementing the decisions.

Naturally, for the system to work, management has to ensure that the workers are enabled, through knowledge and training, and empowered, through being given the responsibility, to take the operational decisions.

This is the model for the fast-moving, flexible corporation of the 1990s and, in so far as they can, potential employees should check out how far their potential employer subscribes to it.

Having looked at how investors, customers and employees should deal with the microelectronics industry, there is one last group of people who need to be considered – the High Street consumers who buy the end product of the collaboration between the microelectronics supplier and the equipment-maker.

You might not think you need to know anything about microelectronics when you go into your local electronics stores to buy a consumer electronics product – but increasingly often you would be wrong.

In many products the value of the microelectronics is becoming much more than the traditional tenth of the value of the end product (in PCs, in the mid-1990s, it represented up to a third) and, when that happens, microelectronics trends control the pricing, cost curve and future direction of those products.

Once products are on the chip price learning curve they should be halving in price about every two years or doubling in capability for the same price.

In practice, equipment manufacturers try to get around this awkward fact, which means that the value of their products is headed inexorably towards giveaway status. The way they do that is by continually looking for upgrades, sometimes rather spurious upgrades.

The industry term for adding a feature or two and a new name to an old product is 'churning'. It allows companies to give the illusion that they are coming out with a continual stream of high-priced products, when in fact the products are slightly revamped old products.

At the same time it gives companies an excuse to take a product off the market when it has got too cheap for them to make the unit margin they would like. By 'churning' they can avoid the effect of the chip price learning curve and keep the prices of end products up.

The answer for consumers is to wait until products are about to be discontinued when they can be bought cheaply and with all the manufacturer's guarantees and support. The High Street stores are usually prepared to give good discounts for products that are about to be superseded by a redesign or new model.

Making sure that you have a guarantee and a helpline is essential unless you are an expert on the products already. Most of the manuals are difficult to follow for the ordinary person and a human contact to see you through any problems is often essential.

However, if a consumer understands the product and is prepared to take the risk of being outside a guarantee or support service – and the more a product is electronics-based the more reliable and long-lasting it tends to be – then it is not a bad strategy to buy electronic goods second-hand.

Exceptionally good prices can be found in the second-hand market and the inherent reliability of electronic goods should make it a safe practice, the reason being that chips should last indefinitely as long as they are kept in reasonable conditions – particularly in a dry environment – so chip-based equipment usually has a long lifetime. This is especially so if it is compared with mechanical equipment.

When buying computers, it is essential to keep in mind that you, the consumer, are interested in applications – not the lump of hardware in the shop. Your first priority is make sure that the computer can run as many applications as you could possibly want.

In the mid-1990s, the vast preponderance of applications were written for the IBM-type of personal computer, which is copied by many manufacturers. All have in common the MS-DOS or Windows operating systems, from Microsoft, and a microprocessor – made either by Intel, AMD or Cyrix – which is known as an 'x86 architecture' microprocessor. If the computer has that combination it will run the great majority of the existing software.

However, for doing creative work, such as design and desktop publishing, the general consensus, in the mid-1990s, is that the other major computer type – the Apple type – is the most appropriate. Apple Macintosh computers have led the way in this type of work. In 1994, Apple started to replace the Apple Macintosh with the 'Power Macintosh'.

And if your only interest in computers is to play games on them, then the Nintendo and Sega brands dominate in the mid-1990s. But the general rule applies – buy the machine that will run the most games applications.

The second half of the 1990s will see a battle develop as IBM and Apple combine to take customers away from the x86 microprocessors of Intel and onto their own proprietary new microprocessor, called the PowerPC. The Power Macintosh is an early example of a computer based on the PowerPC.

The battle is likely to be confusing to consumers – not least because it is going to be IBM and Apple telling people to move to PowerPC-based computers away from the IBM's own Intel-based computers and Apple's own Apple Macintosh computers.

But the basic rule for computer consumers – 'Will it run all my past and likely future applications?' – will remain the best protection for the consumer.

And although salespeople may tell consumers that the computer will run all their programs, the key question to ask is whether those programs will have to run under a technique known as 'emulation'. Emulation involves a program that acts as a translator, allowing a program written for one microprocessor to run on a different microprocessor.

This translation process always results in a performance penalty, which could result in the computer operating five to ten times slower in the mid-1990s, though this could be reduced as emulation techniques are improved.

So the important word to remember when buying a computer in the 1990s is 'native'. Native in the computer world means that the program was written for that particular microprocessor and so will run on it at full speed without emulation. If it is not native, it can only run under emulation and, consequently, with worse performance.

However, the key point for most consumers to remember is that, for many of their applications, old technology is perfectly adequate. For instance, word processing is a very common application and it is difficult to see any difference (in everyday use) in the speed with which the same word processing package is run on the 1970s-vintage 8086 microprocessor and the 1980s-vintage 486 micro-processor, despite there being five generations of technology between them.

So there is a great opportunity for equipment manufacturers to use these very cheap chips from the 1970s to make all sorts of very useful, very cheap equipment. And consumers should never be put off their budgets by slick salespeople saying that they need the latest technology – often they do not.

In the mid-1990s, outsiders are in a better position than ever before to deal fruitfully with microelectronics companies. Investors can expect more information to be disclosed than ever before; customers can expect to get closer to chip-makers than ever before; employees can expect more opportunities for personal advance-ment than ever before; and consumers are realizing that they can expect falling prices.

From practising a black art in near-secrecy, chip-makers are beginning to take as fundamental a role in the modern world as coal miners did in fueling the 19th century industrial revolution, or the oil companies have had in powering 20th century industrial growth.

The effect of these latter-day alchemists on the industrial struc-ture of the world is beginning to have some quite fundamental and startling results – as the next chapter describes.

9

What can we expect?

Change. The high-tech world is changing in the 1990s and changing fast. Across the developed world the great computer companies and telecommunications companies are losing jobs and downsizing at a rapid rate because the pace of technology has overtaken them. They are no longer in control of their business environments.

Take the computer industry. The giant, dominant companies were at their peak in the 1980s – IBM's annual revenues were $60 billion. But they didn't get to be so big because they were super-efficient or super-competitive: they got big because they had total control over every aspect of their product and captive customers.

They not only designed, manufactured and marketed their computers completely by themselves, they also locked customers into their product by making it too expensive for them to switch suppliers. With such total control over their business environment they could charge whatever they liked for their products.

That's why IBM could have layers of management stretching to infinity. Managers managing the work of other managers who managed other managers. Decisions went through innumerable layers of hierarchy. Only a company grown comfortable on a business with fat margins could afford such a structure.

The other big computer companies worked in the same way, maintaining total control over their products, locking in their customers and determining their own profit margins.

Customers were locked into their computer suppliers because the programs that customers bought year after year would only work on one company's computers. To change the computer supplier would be to effectively write off that investment in software. Few purchasing managers dared to take such a decision.

Under a system so controlled by the large companies, it was inevitable that there were rumors of abuses. One well-circulated rumor was that companies would supply computers with some of their functionality switched off. When customers wanted an upgrade, so it was said, the computer company would send round someone to switch on the extra functions... and present an invoice.

It was not surprising that the computer companies could afford layers of managerial hierarchies under such a system. The system rested on one key principle – keep the product proprietary or, in the argot of the industry, maintain architectural control.

So long as a company kept its product proprietary – ring-fenced by the laws protecting patents, copyrights, trademarks and trade secrets – a computer company could ensure that it had no competition to erode its profit margins.

Then along came the microprocessor and changed the rules of the game. Both Apple and IBM made their first personal computers from off-the-shelf microprocessors. In one fell swoop anyone could make a computer. The microprocessor's inventor, Ted Hoff, reckoned 'the microprocessor democratized the computer'.

Not only could anyone make a computer, they could make a computer for a few thousand dollars that was as powerful as some of the computer industry's products that cost tens of thousands of dollars.

One of the problems for IBM when it brought out its first PC was how to hobble its power to stop it being so powerful that it competed with the low-end of the company's range of minicomputers.

At first the microprocessor made possible only the PC, but by the mid-1990s it was beginning to take over all areas of the computer industry as it became possible to bolt together large numbers of microprocessors to make computers of almost infinite power.

That spelt the end of the giant, dominant computer companies with their proprietary architectures and captive customers. They started losing literally billions of dollars in the early 1990s and laid off tens of thousands of staff as they struggled to cope with the radical change in their world.

Eventually, recognizing the culprit behind their problems, the computer companies decided to launch their own microprocessors. IBM and Apple launched the PowerPC, DEC launched a microprocessor called the Alpha, Hewlett-Packard launched one called the PA (Precision Architecture) and Fujitsu and Sun Microsystems launched one called SPARC.

The hope behind these launches was that each of their microprocessors would give each of the computer companies back the proprietary architectures that had been the basis of their fat margins in their heydays. Once again they would have proprietary products into which they could lock their customers and charge them what they liked.

It is likely that they recognized the culprit too late. By the time the computer companies woke up to the fact that the microprocessor was killing their business, one company, Intel, had sewn up 80% of the market for microprocessors in PCs.

Moreover, Intel had become so much the choice of all the PC-cloners over the years that 80% of all the programs written for PCs had been written for Intel's microprocessors.

Naturally, all the software companies had wanted to write for the most widely used microprocessor, because that was the largest market. So over the decade and a half from the first IBM PC all the software companies had written first for the IBM standard based on Intel's microprocessors and only as an afterthought for computers based on other microprocessors.

So at the PC end of the computer business – which represented most of the market for microprocessors – the traditional computer companies bringing out their own microprocessors had the problem of how to get their microprocessors used by PCs that could run useful amounts of software.

Since the amount of software written for Intel microprocessors exceeds that written for other microprocessors by a factor of 100:1 (claims Intel's president Andy Grove), the problem for the makers of alternative microprocessors could be significant.

The other alternative for the computer companies was to accept that their new microprocessors would have to run Intel-type software. It is possible for microprocessors to do that by 'emulation' – a program that will allow one microprocessor to run the programs written for another. However, it comes with a penalty: in the mid-1990s emulation results in five to ten times slower performance for the PC.

That is enough to put most users off – certainly users who want top-of-the-range performance and are buying the latest and most expensive PCs using the latest and most expensive microprocessors. Such people do not want to pay top dollar for the latest machine only to find that it does not run their applications programs any faster because of the 'emulation penalty'.

There are ways to reduce the emulation penalty. One way, for instance, is doing the emulation 'on-chip', i.e. having the emulation software package contained in a bunch of transistors on the chip itself instead of being in a separate software package running off a disk. However, by how much this can reduce the emulation penalty is not, by the mid-1990s, entirely clear.

So all the non-Intel microprocessor architectures either have to come up with a way of running the software written for Intel microprocessors as fast as Intel microprocessors can run it, or they have to come up with a compelling application that will make everyone want to switch to their particular proprietary architecture because it can do something for them that no other architecture can do.

Examples of 'compelling applications' have been around since the beginning of the PC industry. An early example was the spreadsheet called VisiCalc, which was originally only available on the Apple II. PC-buyers, it is said, buy applications, not machines, and the spreadsheet on the Apple II was the reason millions of people bought the Apple II. They bought it simply because a spreadsheet made life easier for them.

This kind of compelling application, which succeeds in selling computers, succeeds because it is a use for computers which no one had associated with computers before. The inventors of the Apple II did not invent it to be a spreadsheet machine but it sold as one.

So, if the computer companies of the 1990s want to come up with a compelling application to try to wrest back control for their own proprietary architectures, the compelling application will have to be something normally unassociated with computers – possibly something in the telecommunications field, like a combined teleconferencing/fax function where people can sit around and talk to each other via PC screens and physically pass documents around.

And when it is remembered that the 1990s is the age of digitization, in which every form of information – data, TV and audio broadcasting, video, speech, graphics and photographs – can be sent and received in the form of 0s and 1s, it is clear that the potential for adding many more functions to the PC is enormous.

So, if computer companies want PC-buyers to move to PCs using their microprocessors they must either invent some compelling new application, or they must make their microprocessors run all the software Intel microprocessors run and run it just as efficiently.

If they can do neither of those things, then their only course of action is to start getting used to the margins earned by PC-makers and start shedding those corporate layers of hierarchy.

The frightening thing for the computer companies is that the trend from proprietary products to products based on off-the-shelf microprocessors won't just stop at the PC level. Their entire businesses are under threat because not only PCs, but also large computers, are nowadays being designed using commercial microprocessors.

Microprocessor inventor Ted Hoff's prediction that the microprocessor would 'democratize the computer' will be fulfilled when all computers, from PCs to supercomputers are made from off-the-shelf microprocessors available to anyone. Already that is happening in the mid-1990s.

When all computers are made up of banks of microprocessors, the company that controls the dominant microprocessor architecture could end up controlling not only the PC business, but also the computer industry. No wonder the computer companies are coming out with their own microprocessors!

The problem the computer companies now face in trying to establish their own microprocessors is not simply the software compatibility issue, but also the problem of how to support a microprocessor development program financially.

This can be horrendously expensive. Intel, in the early and mid-1990s, consistently spent two billion dollars a year on developing new microprocessors and building the factories to manufacture them. But whereas Intel, the reigning cock of the roost in microprocessors at the time, had a large, very profitable revenue stream from its current range of microprocessors, rivals trying to establish new microprocessors did not enjoy such an advantage.

So the computer companies, without microprocessor revenues of their own, will have to use revenues generated elsewhere in their businesses to keep up with Intel in bringing out new generations of ever more powerful microprocessors.

And the revenues generated elsewhere in the computer companies' businesses are shrinking both overall and in profitability. Keeping up in microprocessors could start to become a strain even for them.

One way for them to defray those costs is to sell their microprocessors to other companies wanting to make clone computers. But whether or not clone-makers will buy rival micro-processors in large numbers, knowing they may not be able to run the majority of the applications as fast as Intel microprocessors, is unknown.

So, in the mid-1990s, control of the computer industry could be shifting fundamentally away from the big computer companies that used to do everything – design and make the computer, manufacture all the components, write all the software and do all the sales and marketing – to an industry where there will be separate companies for each field: design, microprocessors, software and distribution.

Companies in each of these fields will be trying to maximize their own margins as they wrestle to gain some advantage over the companies in the other fields. But the microprocessor-makers will always have the built-in advantage of controlling the future direc-tion of the final product – the computer.

This revolution overtaking the traditional computer industry has been caused by the microprocessor. Invented in 1971, it is only in the 1990s fulfilling its potential as the defining force in the PC industry. To the traditional equipment-makers it is a rude shock that a component can become so important that it controls and defines their business environment. But that's exactly the way things in the computer industry are turning out to be.

However, in the mid-1990s there is an imbalance in the micro-processor field, because Intel has 80% of the market. Not only is this bad news for the large computer companies seeking to estab-lish their own microprocessors but it is also bad news for the PC-makers, because Intel has used its powerful position to charge very high prices for microprocessors.

For instance a top-end PC based on the then top-end Intel micro, the Pentium, cost around $3000 at the beginning of 1994, and the Pentium microprocessor inside it cost $1000. That's a nice margin for Intel and a poor one for the clone-maker.

Thousand dollar microprocessors used to be an obscenity – the sort of price you could only charge the military. With the costs of making a silicon wafer in the low hundreds of dollars, and the prospect of getting 100+ microprocessor chips off a single wafer, it is not difficult to see why Intel was judged, in a 1994 article in *Fortune* magazine, to be the 10th most profitable large company in America.

For the poor, margin-starved PC-maker, there is some hope of liberation from the thousand buck microprocessor. That's because rival makers of Intel-type microprocessors, of which the only significant ones in the mid-1990s are Advanced Micro Devices and Cyrix, started to win legal victories that would allow them to make their versions of Intel's microprocessors, and companies like DEC, IBM and SGS-Thomson started to make their factories available to the designers of Intel-type microprocessors.

Once there is significant competition in the market for Intel-type microprocessors, prices will drop like a stone. That means that the price for the latest PCs will also drop like a stone, because there is no dominant PC-supplier who can afford to start trying to claw back the margin from the microprocessor to the PC.

Fortunately, there are so many PC clone-makers out there that no PC-maker is going to become dominant and so be able to start dictating computer margins again. Meanwhile, Intel's response, in the mid-1990s, was to bring out its new microprocessors at ever-decreasing intervals, so accelerating the pace of technological development in the hope of burning off potential rivals.

So the happy prospect looms of ever-increasing microprocessor performance leading to ever-increasing PC capability at ever-decreasing prices.

That's at the top end of the business dealing with the latest computers using the newest microprocessors for running the most advanced software and the most recent additional functions such as video-conferencing, compact discs and video.

Of course, if users need a PC for simple tasks like word processing, they do not need a Pentium, or anything like it – a 1978 vintage 8086 will do the job perfectly adequately. For such people, extremely cheap machines – to the point of giveaways – will become available as the chip price learning curve takes its inexorable effect.

So, the mid-1990s scenario for the computer-buyer looks promising. More people are making Intel-type microprocessors, more alternatives to Intel-type microprocessors are being designed, and Intel is bringing forward its plans for more and more powerful microprocessors. All of which means that technological advance will accelerate while prices drop.

The more microprocessor-makers and computer-makers there are, the better it is for the consumer. Only too often, for instance with the traditional computer companies, we have seen that when big companies control an industry, innovation slows down. Big companies don't want innovation if they can keep milking profits from old technology. So for the 1990s, the scenario looks good with the underlying dynamics of the industry forcing the technological pace.

The story of job losses, downsizing and loss of control by the dominant computer companies applies equally to the telecommunications industry. There are two main forces driving change: new technologies and government deregulation.

The two new technologies, which are challenging the old telecommunications technology of copper cables buried in the ground, are optical cables (using flashes of light instead of pulses of electricity) and mobile phones using radio communications through the air.

Government deregulation, which is happening in some developed countries, is replacing the state-owned telecommunications monopolies with a number of new, private companies providing competitive telecommunications services.

Once governments in the developed world started allowing anyone to apply for a license to run these services, growing competition in the 1990s began to erode the businesses of the old national telecommunications monopolies like AT&T in the USA, NTT in Japan, the Bundespost in Germany and British Telecom. Substantial downsizing and job losses are the continuing result in the mid-1990s.

As licenses have been granted, an assortment of organizations, usually consortia, have started building the infrastructures for both optical cables, now colloquially called data superhighways, and for pocket telephones.

Unexpected organizations have become involved: for instance the canal authorities, who see the potential of using their network of towpaths as good places to bury these optical cables, or electricity companies, who can string the cables from their pylons, or water companies, who can stick them down the drains, or railway companies, who can lay them alongside their railway tracks. Other companies are digging up pavements to bring these superhighways to the home.

What cable delivers is very many more telephone calls, or very much more of any other form of digital information, than a copper cable can – to the extent of a 100 000 times increase.

For instance, an optical fiber as thin as a human hair can carry 300 000 simultaneous phone calls at a transmission rate of 2.5

gigabits per second. Furthermore, an optical cable will contain a number of such fibers, and the transmission rates are getting faster: 10 gigabits per second cable should be being installed during this decade.

The capability is mind-boggling – if you take a 10 gigabits per second cable, containing ten separate fibers, then one cable will be capable of handling twelve million simultaneous phone calls. The Americans have a buzz-phrase for the capability – 'bandwidth on demand' – which is generally taken to mean constantly available connection via the phone line at little or no cost.

The beauty of optical cable is that, once it is laid, the use of the cable costs virtually nothing. So, clearly, the traditional phone companies are on a hiding to nothing if they try to compete with superhighways – unless they themselves adopt superhighways.

This is one reason why the traditional phone companies are shedding staff and downsizing their operations. And it is also a reason why the price for using the telephone network should collapse dramatically in the 1990s.

However, the superhighway is capable of doing a lot more than carrying phone calls. Any information that can be put into the 1s and 0s of digital language can travel on the superhighway. And that means practically any kind of information you can think of.

Even a kiss. For instance imagine being away from home. You talk to your partner back home and then kiss an electronic device which records your lip-print, pressure etc., digitizes it and sends it off down the superhighway where it can be reproduced on a similar device at home.

It's much the same principle as described in Bell's patent, except that it's a vastly more complicated message than any he envisaged. So huge is the potential for sending every conceivable type of information that people think that most of the uses to which we will put the superhighway have not yet even been imagined.

They say it is as significant an advance in communications as the canals, or the railways, or the car, or the telephone, radio, TV or aeroplane were. And they reckon that superhighways will have as dramatic an effect on creating new industries and activities as they all did.

Just as the aeroplane was first exploited commercially for its curiosity value in flying circuses and then spawned the modern mass travel, hotel and tourist industries, so the superhighway could spawn similar effects, it is claimed, if we are imaginative enough to think up new uses for it.

It is fun to try. The obvious things are being able to access almost any TV station in the world; but, because the superhighway is two-way, you'll also be able to do things like connecting up to the supermarket, seeing what's on offer and what the price is, putting in an order, paying and giving delivery instructions – all over the phone; or you'll be able to dial up a book or film or an old TV program; or see who you're talking to on a video-phone; or check your bank balance and make transfers from one account to another; or ring up the travel agent and get some videos of holiday locations zapped down the line to you.

Getting even more imaginative, you can envisage contacting the local estate agents and getting them to put on-line all the houses they have for sale and then, donning a 'Virtual Reality' helmet, you could 'walk' through all the rooms of a house.

Virtual Reality is something that the superhighway should make into a huge new industry. It currently involves putting on a helmet in which there are a couple of screens, one for each eye, to which video signals are sent which allow you to see a different part of a view if you turn your head or move it up and down.

Additionally, Virtual Reality suits are being developed that can transmit sensations of heat or cold, of wind and sunshine, of touch and feel. Maybe taste and smell will be added. Such suits would allow you the illusion of 'touching' people in a film or someone on the end of a videophone.

However, Virtual Reality is already exceeding the limits of the helmet and being incorporated into the gyroscope. Imagine a room like an airline pilot's flight simulator in which all the walls, the ceiling and the roof are screens and which is mounted on gyros to give movement. The experience of a fighter pilot, a racing driver or a roller-coaster could be simulated.

And what's more, because the superhighway is two-way, you'll be able to control your environment. As the fighter pilot, you'll make turns and climbs which lead to disaster or survival. On a Virtual Reality cruise, if you got bored, you could have the cruise ship sink.

Of course, the amount of digital information needed to reproduce these sights and sensations would be vast – but so vast is the capacity of the superhighway that it could handle it.

Maybe it would not be necessary to buy the travel agents' holidays. You could have a complete reproduction of a holiday destination delivered down the superhighway to your armchair. No need to go to the airport, sit in a badly air-conditioned metal tube and suffer jet lag – unless of course you like those experiences, in which case they could also be fed to you digitally, down the superhighway, into your Virtual Reality world.

Of course, all these possibilities open up the prospects for new industries. All this Virtual Reality kit will have to be made by someone. In the mid-1990s Virtual Reality systems are so expensive that you have to travel to an amusement center or a defense/aerospace manufacturer to see them in action.

Amusement centers use them for simulating racing drivers' or pilots' experiences and can defray the heavy cost of the equipment by charging people to use it by the minute. Defense and aerospace manufacturers use them to design complicated systems in three dimensions.

However, the inexorable rule of the industry is that what costs a fortune in one generation becomes a commodity in the next. Once these Virtual Reality products become subject to competition and the chip industry's learning curve, they will rapidly erode in price to become affordable in the home.

The opportunities for the manufacturers of Virtual Reality hardware will be exceeded by the opportunities for the creators of software. Those people who can think up games, services and diversions to send down the superhighway will command unlimited business opportunities.

Anyone who can think of information, entertainment or sensations that people want to buy will be able to sell it to a superhighway services provider – just as independent program makers sell their wares to TV stations. That will give enormous opportunities to the creative and the imaginative.

For the business world, the prospect of the ever-open superhighway link between foreign offices and factories provides staggering opportunities. Company meetings will be simple to handle between people on each of the five continents simultaneously. Discussions, document exchange and amendment, approval of technical drawings and plans and signed authorizations will be possible down the superhighway.

Already multinational companies often conceive, design, plan, manufacture and test a product on five different continents –

choosing each one for the particular advantage that region can bring to a particular aspect of a project. To coordinate such a project, video-conferencing links are maintained.

As the superhighway comes into its own, such cross-continental projects will become open to all companies – not just large ones – significantly stimulating opportunities for international business.

Similarly, if you want to sell your product to a company on the other side of the world you'll be able to make a presentation via the superhighway. Your customers will be able to ask questions, have plans and graphics sent down the superhighway to them, be able to change these plans and send them back, and be able to exchange contracts and make payments.

The other possibilities of this ever-available, inexpensive link will be in the field of 'narrowcasting', as opposed to broadcasting. Narrowcasting means there is no need for anyone in the world to be excluded from any occasion in the world except on the grounds of privacy. For instance, say you are a student at Oxford University and you hear that a professor at Harvard University is giving a lecture on your subject.

You ask Harvard for permission to 'attend' electronically, maybe pay a fee, and you get linked into the lecture. It would be an attractive way for professors to augment their salaries and, with the plummeting costs of video cameras and potentially inexpensive use of the superhighway, the mechanics of setting up the link should become affordable even to students.

There are as many examples of narrowcasting as there are minority interests in the world. A medical student could be linked in live to watch an operation on another continent; a top specialist surgeon could attend a difficult operation in another country as an advisor; a group of stamp collectors in Tokyo could be electronically joined by a philatelist from Toronto; a new car being launched in Detroit could be inspected by a prospective distributor in Seoul; if you could not attend a party you wanted to go to, you could have it linked up to your home.

The possibilities for narrowcasting are pretty well endless. One that has been put up by the Massachusetts Institute of Technology is that you have your own special newspaper – tailored to all the subjects which you find especially interesting – delivered by the superhighway to your home every morning, where it is then printed out on paper used by the previous day's newspaper, which has been through your own paper recycling machine in the meantime.

Also, the prospects for surveillance become almost limitless. With video cameras reducing in price and linked to the superhighway, it is not fanciful to expect that we shall see much of the inhabited world permanently on camera. To the extent that this controls crime, so much the better; but to the extent that it allows the unscrupulous to impose and sustain tyrannies, so much the worse.

So the general effect of the superhighway on the telecommunications industry will be take telecommunications out of the hands of the traditional telecommunications companies and give it to the new companies installing the superhighways.

That will make for a highly competitive environment for providing the means of communication, which in turn should mean sharply reducing prices, so giving the information providers an inexpensive vehicle for delivering their products.

The expectation is that the profit will be made not, as now, from charging for the time spent using the line, but from charging the information provider for access to the line who, in turn, will charge the information receiver for the information ordered.

The expected diminishing returns from charging for time on the line are propelling the traditional telecommunications network operators to get into the business of providing entertainment and other services. In some cases, e.g. in the UK, there are government restrictions on their doing that.

So that is how optical cable, or the superhighway, is threatening the traditional telecommunications companies. Governments will most likely continue to license more providers of services and entertainment, and so, instead of national monopolies, we will get a growing number of companies competing to provide services down the superhighway.

The other technological threat to the traditional structure of the telecommunications industry is the growth of pocket phones. Again governments have licensed growing numbers of different companies to set up the networks for these services and more will undoubtedly be licensed in the future. The more there are, the more competitive the business environment for the network owners will be. Consequently the pricing for airtime will fall.

Simultaneously with the falling cost of airtime, the cost of the pocket phones themselves should be falling substantially. The chip industry learning curve is remorselessly beginning to erode the cost of the phones.

In the mid-1990s these can be made with about half a dozen chips, but we can expect that number to halve every three to four years according to microelectronics industry tradition. So it's a safe prediction that the declining cost of phones will reduce them to the level of cheap consumer goods before the end of the 1990s.

The size and weight of the phones should also be reducing rapidly. As the transistors shrink and so reduce the number of chips needed to make a pocket phone, the additional benefit is generated of a lower power requirement. That's because the smaller the transistors, the less power they need to operate.

A lower power requirement means batteries need to be less powerful. That means the batteries can be smaller. As the main contributor to the weight and bulkiness of pocket phones, the shrinking battery pack will be the biggest influence on miniaturizing portable phones.

With airtime cost falling, and the cost of pocket phones falling, the prospects for wireless communications look good as more people get interested in providing networks and get licensed to do so.

In the mid-1990s, the networks are all of the 'cellular' variety where the call is routed to a transmitter and then into the telephone system. However, although cellular is suited to highly populated areas, the economics of setting up and maintaining a network of transmitters makes cellular less attractive in less densely populated areas.

Accordingly, it is thought unlikely that any more than 20% of the world's landmass will be covered by cellular networks by the year 2000.

For the open spaces of the less populated areas, the future of wireless phone networks lies with satellite-based systems. By the late 1990s, it is expected that the first satellite systems – whereby calls are routed across the world via a string of low or intermediate orbit satellites – will be in place and working.

Two consortia, one led by the international maritime navigational organization Inmarsat, and the other called Iridium, led by the US company Motorola, are intending to have satellite systems up and running by the late 1990s. Five other organizations, all American, are looking at establishing satellite networks: TRW, Ellipsat, Constellation, Loral/Qualcomm and Teledesic – a joint venture between Bill Gates of Microsoft and Craig McCaw of McCaw Cellular.

The potential for the networks is enormous. Even in an exploding market the number of expected owners of pocket phones is currently only expected to be 150 million by the year 2000 – insignificant compared with the 5.6 billion global population.

As with the data superhighways, establishing the networks is only the beginning. Once up and running, hardware companies will be making (and are already making in the case of cellular networks) equipment to hook onto the network – things like portable fax machines, pagers, email receivers and senders, and video cameras; and the software providers will be thinking up services to provide to the global nomads using these wireless networks.

An obvious example of a service for the nomads is the opportunity to receive their regular daily newspapers anywhere in the world – beamed by a satellite to a traveler's flat-screen reader plugged into a mobile phone in, say, Outer Mongolia, maybe even before it hits the breakfast tables of its country of origin.

Other services that global wireless networks could provide are travel advice, local contact names, local news and local maps in the traveler's own language and a menu of 'help' routines in the event of trouble.

So in three main ways the advent of new technologies like superhighways and pocket phones on wireless networks provides new business opportunities. First, it allows new people to play the network operator role, providing competitive networks and so reducing costs; second, it is a catalyst for industries building products that people buy to hook onto the network; and third, it gives a major opportunity to the providers of services on the networks.

Change in the high-tech industries is not confined to the computer and telecommunications industry. It is also coming to the broadcasters, and for much the same reason – digitization.

With Japan announcing in 1994 that it was going to back digital TV broadcasting, the developed world is now agreed that broadcasting's future direction is to be digital, with digital TVs receiving digital broadcast signals. Such broadcasts are already beginning in much of the developed world and will gradually take over from analog broadcasting.

This means that in computing, in telecommunications and in broadcasting the technology for transmitting, storing, displaying, processing and accessing information is going to be the same – the 1s and 0s of digital language.

The computer gets its stream of 1s and 0s from a magnetic disk, the telephone gets its digital stream down a wire or optical cable, and the TV gets its digital stream through the air.

So digitization will mean that it will be a simple matter to merge the functions of the computer, the telephone and the TV. It will also mean that the entertainment you receive on the merged computer/telephone/TV will be sent in three ways: down a phone line, from a disk or via a broadcast.

And new entertainment providers will be looking to supply their wares down all three media. Which means that the established television broadcasting companies face the same problem as the traditional computer and telecommunications companies – new players – and with new players, fiercer competition.

And when the satellite TV broadcasters are added into the equation, the broadcasting business looks less and less like the highly regulated license to print money of pre-1990s days, and more and more like a highly competitive, cut-throat commercial industry.

The good news for us in all of this is that first we get increasing choice of services at ever decreasing prices as the competition to run services hots up, and secondly, when the equipment we need to receive these services goes digital, it will be firmly set on the microelectronics industry's price erosion path.

Once the electronic guts of each piece of equipment is reduced to digital chips, the remorseless shrinking of the transistors will ensure that the equipment gets cheaper and smaller every year, and will result in many different equipment types merging into one all-purpose tool: faxes, copiers, printers, video recorders etc.

So, digitization is providing both a radical change in the equipment and a revolution in the structure of the industry. In the three largest industries spawned by electronics – computing, telecommunications and broadcasting – the big, traditionally dominant, players are having to respond painfully to their rapidly changing environments – principally by shedding staff in very large numbers while downsizing and refocusing their operations,

It is much the same story in another area of the electronics industry – the defense industry. With the end of the Soviet military threat, Western governments have cut back on their defense R&D and procurement budgets.

In a symbolic expression of this President Clinton dropped, in 1993, the word 'Defense' from the US Defense Advanced Projects Agency – so renaming it ARPA instead of DARPA – which is one of the leading US government bodies for funneling money into R&D. The intention was clearly signaled that the focus of government research money could be switched to more commercial objectives.

And in 1994, Clinton announced a billion dollar research program via ARPA with the aim of establishing a manufacturing industry in the USA for flat panel displays. That an American government should directly intervene to support manufacturing industry was seen as something of a novelty – suggesting a different approach to the government/industry relationship.

Certainly the USA is concerned at its growing deficit in the trade balance on electronic goods, and the end of the Cold War gives it the chance to devote more resources to industrial areas in which it has lost its technical lead.

Also there are signs, in the mid-1990s, that the European Union is beginning to accept that it needs to get more directly involved in supporting manufacturing industry. The European deficit in electronics manufacturing, which had been growing steadily during the 1990s – was one reason to concentrate the minds of European leaders; another was its 19 million unemployed people.

At the same time, Europe appeared to be taking on board the lesson learned by America during the early 1990s recession – that in the 1990s it is small and medium-sized firms that create jobs, while the big companies tend to shed jobs – even in the high-tech area.

For Europe to start thinking like that was a considerable reversal of the thinking throughout most of the second half of the 20th century. For most of that time the thinking was to promote 'National Champions' – large companies in each of the European countries that received the bulk of government financial support for high technology.

In the mid-1990s, with the trade imbalances in their electronics sectors large and growing, America and Europe will be tempted to switch more of their R&D budgets from military to civil research and to start actively supporting their manufacturing industries. Such moves could start to redress the long relative decline of the West's competitiveness compared with Asia.

So, for entrepreneurs, the late 1990s should be a time of opportunity. Governments are looking kindly on them as job providers; the new optical and wireless networks give the chance for providing new services; the big companies no longer control the high-tech environment and are often uncertain of their future direction; and fast-changing microelectronics technology ensures that the high-tech scene is in constant change.

Although the entrepreneurs and companies that are going to be successful in the second half of the 1990s will have to understand these changes, they will need something else that is even more important – imagination – and that is the subject of the next chapter.

10

Imagination

This is the chapter you should be writing. Your imagination is probably better than the authors', and, once you have a basic frame of reference about the nature and capabilities of microelectronics, the rest is up to the imagination.

The great product breakthroughs in the electronics industry have come not from the traditional companies but from the unconventional, often the anarchic, usually the entrepreneurial: take Edison with his light bulb, gramophone, and microphone; or Morita with his Walkman; or America's remarkable breed of creative technocrats who pioneered the personal computer.

The 'techies' first gathered in 1970, when Xerox Corporation, the copier company, swallowed one of the myths of the age – that computer technology would result in the 'paperless office'.

The idea was that, with most large companies adopting computers, everyone would get their information from screens linked to the company mainframe, rather than from bits of paper.

To a company that made its money from copying bits of paper, the vision was appalling. An alarmed Xerox decided that it needed to become a big player in the upcoming era of the paperless office by supplying all the new kinds of equipment that would be needed for it.

The company began by hiring fifty of the best computer researchers it could find in universities and industry and installed them in a building in Palo Alto which it called the Xerox Palo Alto Research Center or, as it became famously known, Xerox PARC.

By 1973, a mere three years after it was set up, the PARC team of iconoclastic, independent-minded, free-thinking techies had invented the personal computer, the laser printer, the high-speed computer network and the first easy-to-use computers using 'graphical displays' – where you point to a picture of what you want rather than type in commands.

So why doesn't Xerox dominate the PC business? The answer is because it didn't make, let alone market, any of these things. To the disgust of the PARC technologists, Xerox management walked away from exploiting the developments.

It took another ten years for the PARC ideas to be implemented. And it took another bunch of similarly anarchic technocrats to do it. They were the techies of Apple Computer, founded by a couple of students, Steve Wozniak and Steve Jobs, in 1977.

In 1979, Jobs had visited Xerox PARC and immediately recognized that it had defined the future direction of computing. In 1983 and 1984, two computers from Apple incorporated many of the PARC developments – first the Lisa, which, at $10 000, was a flop, then the $3000 Apple Macintosh, which was a wild success.

In its early days Apple, like PARC, was a place where creative technocrats held sway deciding things through brainstorming sessions. So unstructured was the company that it apparently ran for its first five years without any budget. So much money poured in that no one thought to check what was spent.

Even in the 1990s, much of the personal computer industry's early culture lives on, with the people driving the new ideas still tending to be creative techies. So it is not surprising that the industry is characterized by constant change.

And as consumer electronics and telecommunications all merge into and onto the personal computer, that means the largest chunk of the electronics industry is in a state of change – still up for grabs by whoever can bring the most fertile imagination to bear on its direction.

The greatest leaps in product innovation have often come from the contrary thinkers – some of them without any formal training in those academic disciplines – electronic engineering, computer science, physics, mathematics, software sciences etc. – that relate to the industry.

The only vital starting point to imagining the products of the future is a knowledge of the technology's present capabilities, future capabilities, its likely cost projections and an understanding of what amuses, helps, teaches, interests and stimulates the human being. By putting those elements together, the great blockbuster products of the 1990s and the 21st century will be invented.

The purpose of this book is to provide that starting point so that as many minds as possible can be brought to bear on dreaming up the products of the late 1990s and early 2000s.

The starting point is based on the fastest advancing electronic technologies of the mid-1990s – the memory chips, the microprocessors, the programmable logic chips, magnetic disks, optical disks, high-capacity interactive cables, cheap and tiny video cameras, cheaper flat screens etc.

These are the fast advancing building blocks that will be put together in any way the human imagination can dream up to create the products of the future. They will be put together to provide products in which any or all of a host of different information types – written words, spoken words, music, still photos, videos, films, broadcasts, narrowcasts, drawings etc. – will be stored, received, transmitted, compared, analysed or sifted through.

Breathing life into these hardware products will be all the people who can create the uses to which these products will be put: games, learning aids, demos, Virtual Reality experiences, conferencing, entertaining, work aids etc.

There will be an increasing need for creative people who can think of ways of packaging up all the many information types in the best way to make an easy-to-learn or easy-on-the-eye or persuasive or simply entertaining program to run on the hardware.

The new technological potential is so open-ended that even the smartest companies and people have little idea of the sort of hardware products people are going to want, or the sort of programs people will buy to run on them.

Since people are generally unaware of the power of the technologies, they won't even have begun to think of the possibilities of what they might want to do with them. So it's a fair bet that most of the products of the age of virtually free storage, transmission and processing of all these types of information have yet to be thought up.

At the moment we tend to see the new technological capabilities as something that provides different ways of doing what we already do. For instance, people are already digitizing books and films so that they can be squirted down phone lines or delivered in the form of CDs. And in the mid-1990s people went round buying up the rights to digitize famous paintings in art galleries.

The idea is that you can call up, from an optical disk or from a picture bank via a phone line, the digitized paintings and display them on-screen. Indeed, with dozens of the new flat panel screens hanging on the wall, maybe all in nice gilt frames, you could have a constantly changing picture gallery – Impressionists one day, Dutch Masters the next.

That's the kind of new use for technology that people thought up when the early cars replacing horse-drawn carriages were built to look just like horse-drawn carriages. When aeroplanes were first invented, their early commercial uses were for flying circuses – wing-walking and the like – as a suitable way to make money from a curiosity.

Similarly, when the personal computer came in it was thought of simply as a smaller large computer – i.e. something in which you stored information. It was only when people started thinking up spreadsheet programs and word processing programs and computer graphics and games and the like that people started seeing the personal computer as something completely different from a mainframe computer.

The lesson seems to be that changing technology runs far ahead of people's ability to use it or even see a use for it. Human nature, being naturally conservative, naturally disliking change, naturally clinging to the familiar, is the main barrier to the use and adoption of technological progress.

Yet the world is faced, in the 1990s, with so many apparently insoluble and potentially apocalyptic problems that it needs to use every tool at its disposal to try to find solutions. Very cheap, very accessible technology is one of the most powerful tools it has.

For instance, one of the besetting problems of the world is unemployment. And one of its causes is lack of education and training. Yet we can now create, via the merging of so many different information types, the most powerful and accessible and interesting training programs in history.

Education systems around the world are groaning for lack of funding and shortage of teachers. It will be much cheaper to provide entertaining, high-quality learning programs for kids via cable and optical disks than by hiring more teachers. And learning via the screen can be much more personalized, enthralling and rewarding than that delivered by a harassed teacher to over-populated classes.

And with the virtually free transmission of information that is coming in the 1990s, it should be possible to deliver education and training globally, instantly, to millions of people at negligible cost.

Take another problem – the world is short of people to write software. The world has always been short of people to write software. And with the explosion in the power of computers, which means that more and more powerful software will be required, the world is probably always going to be short of people to write software.

Therefore, globally available programs to teach people how to write software would benefit the world, first by balancing the demand and supply of software; second by bringing better paid work into poorer countries; and third by bringing to bear more good brains on the problem. There is no regional monopoly of human talent – the more good brains that are tapped throughout the world, the better the world's work will be done.

In many less intellectually demanding areas than the writing of software, the richness of new training aids – with sight, sound, pictures, video etc all combined – should be able to make many jobs and disciplines accessible to many people who might not have the opportunity to train in the normal way.

If people are not to have work, then they should have diversion. The peculiar achievement of the governments of much of the developed world in the 1990s is that they created mass alienation of large numbers of citizens at a time when mass communication has never been more universal and more inexpensive.

In the 1990s, using high-speed (2 megabit per second plus) 'information superhighways' or using high-capacity optical disks that each can store 5000 megabits of information (i.e. five gigabits or 640 megabytes) we have the basic, inexpensive, universal means to train and entertain those who want to be trained or to be entertained.

Having mentioned the optical disk, it is probably necessary to give an account of this still evolving technology. The optical disk, sometimes called the video disk, laser disk, compact disc or CD-ROM, was first used for music and, in that context, is familiar to everyone as the CD.

However optical disks can now do much more than store pre-recorded music – they can store every type of digital information – video, speech, text, graphics, sound and pictures. They are able to store large amounts of information – about five times that of a hard disk in the mid-1990s. They can currently hold up to around 800 megabytes of storage – enough to hold a couple of 90 minute compressed videos – at a squeeze – or a vast amount of written data – e.g. 400 copies of *Gone with the Wind* (or 1200 copies if compressed).

Early optical disks could not be recorded by the customer. In the mid-1990s, optical disks became available that could be recorded by the customer – but they still couldn't be erased and re-recorded.

The storage capacity is so great that the prospects of putting every type of training course – combining pictures, words and speech and even setting tests – on optical disks is being undertaken.

Dictionaries on optical disks are commonplace. Putting all the back issues of a newspaper or magazine on optical disks is also being done. Optical disks provide the potential for everyone to have access to the sum of human knowledge.

Another kind of optical disk that is emerging is called magneto-optical or MO for short. MO is a little way behind optical disks in terms of storage capability, but is catching up fast, at around 500 megabytes in the mid-1990s. It has the great advantage of delivering its information about ten times faster than a CD-ROM and the even greater advantage of being erasable and rewritable.

MO disks are the same size as 3.5 inch floppies and provide the prospect of carrying around large amounts of self-recorded, updatable, information.

So the optical disk is providing a method of storing every type of digitized information in vastly increased quantities, while the optical cable provides the means of transmitting it in vastly increased quantities.

The dilemma for the world's governments will be whether to try to influence the use of these powerful tools in the same way as the great public libraries of the 19th century were set up – as a way to provide education, enlightenment and civilized diversion to the many at no cost – or whether they will be set up for the benefit of tycoons who will use them to deliver whatever they can charge the most money for – usually sex and violence.

There can be little doubt that the world is communication-starved. How often in our entire lives do we have conversations with people from different cultures? Apart from the odd encounter on holiday or traveling for business, it is rare. And it is rare for a good reason – different languages and the expense of foreign phone calls.

Microelectronics in the 1990s should put an end to both of those constraints. By the end of the decade the chip technology should be there to make it cheap enough to incorporate a translating device into every phone line. And by then the cost of international phone calls should be low.

To be able to see and hear and talk to and understand another person from a different culture has far more impact than reading about another culture, and is an infinitely more accurate portrayal than watching a TV series such as *Dallas*.

Making person-to-person contact possible and inexpensive should be a major catalyst to creating, if not a world culture, then at least a world in which different cultures understand and tolerate each other.

For world trade, the arrival of cheap communications without language problems could be a sensational catalyst. With deals being struck and documents being exchanged instantly, commerce could see explosive expansion.

And if the developed world is lacking in training, education, professional expertise and know-how – then so much more is the developing world. The limitless aspirations of the Third World could start to begin to be met by the inexpensive provision of understandable information.

In many more mundane ways we are communications-starved. How many jobs for people with your skills are vacant in your county, your city, your country or even in the world? Few people know. But with the information superhighway connected up to your television/PC, it will be possible to find out.

How often do you have a phone conversation with more than one person? For most of us, 'conference calls' at work are the only time – hardly ever do we do it for fun.

However, in the real world, when we go out for our recreation to the pub or for a meal, the pleasure derives from talking to and listening to several people. The constraints of telephones have artificially limited us to one-on-one telephone conversations.

In the 1990s, the plummeting cost of phone calls and the increasing sophistication of telephones should make it possible for us to do what comes more naturally – mix as a group electronically – simultaneously talking to and seeing a group of people via the superhighway.

Another instance of communications starvation is meeting people. For instance, how many members of the opposite sex do you meet before you marry or co-habit? 100 perhaps? How many do you get to know well (e.g. through half a dozen longish conversations)? A dozen perhaps?

Well, the data superhighway could link you to banks of pre-recorded spoken and video messages from other people and, if you liked someone, could fix you up with a live sight-and-sound electronic chat.

That's a far less emotionally taxing way of meeting someone than a blind date, and, judging by the exploding columns of 'Lonely hearts' advertisements in the world's newspapers and magazines, one that would be increasingly popular.

A good many people like the sense of distance which doing something electronically, rather than face-to-face, allows. Human contact has its terrors for many – even, it appears, in something so apparently mechanical and emotionless as cashing a check in a bank.

When the first hole-in-the-wall, automatic cash dispensers appeared on the streets, it was noted by the banks that people went to them twice as often as they used to go to the bank to cash cheques, and each time they went they took out half as much money as they used to.

That's a pretty sure sign that they preferred the mechanical bank clerk to the human one. Maybe it's less embarrassing to be told you have no money in your account by a machine than by a bank clerk.

For the same sort of reasons, many of us are put off going to see the dentist, the doctor, the accountant, the bank manager or the lawyer – though judging by the demand for expert advice through the public help services there is a need for more such advice than is being supplied.

With virtually free information storage it is possible to store the knowledge of an expert – and all the ways in which that knowledge is interpreted and applied – for almost nothing. And with virtually free transmission, it will be possible to squirt that knowledge down the phone line into your home computer, or into a storage device in your TV, in a fraction of a second.

It is not only satisfying to think that the entire professional knowledge of your accountant can be squirted down a phone line in a few milliseconds, but it should also empower a lot of people to set up businesses or seek new careers or get a loan to buy a new house or get rid of a nagging ache or pain.

In the mid-1990s, banks are already coming out with do-it-yourself loan vetting programs where you input your personal details and out comes a decision. There seems no reason why most financial services cannot follow suit.

However, there is a pre-requisite to making all these things happen – a favorable government regulatory regime. Already many

services – like global wireless telecommunications and data
communications – have been delayed in the early 1990s while
governments argue over the technical standards that should be
adopted. So frustrated have technology companies become that
they tend to think 'any standard is better than none'.

Sometimes governments delay in the hope of getting their local
industry's standard accepted and sometimes because they know
their local industry is not ready to take advantage of the market
whereas another country's industry is ready. Sometimes it is
genuinely better to wait for a *de facto* standard to emerge in the
market.

More sinister reasons for delay are because governments are
terrified that systems will not be 'buggable', i.e. they will be totally
secure to the users. Governments around the world think it necessary that they have the ability to bug and can hold up licensing
until they get it.

Another reason for delay is when countries fail to agree on key
standards for actually working the equipment – such as the radio
frequencies that will be allotted to services such as pocket phones.
This has held up the development of markets because it makes
manufacturers scared to come in with equipment working at one
frequency in case it becomes unsalable as another frequency
becomes the standard.

Another problem for providing universal, cheap, knowledge-
intensive services to the people who most need them is that
governments may be influenced by the network owners and service
providers to license very few of them – so allowing them to charge
more – or only to make the networks and services available to
richer neighborhoods, and not to poor ones.

It will take firm regulatory measures by the authorities around
the world if there is to be the necessary fierce competition to
provide networks and services on such a basis that the networks
and services become cheap and accessible to all.

It will also take firm action by governments to make sure that the
right kind of information is available on the networks. All sorts of
professional bodies and vested interests may object to very cheap
financial, legal and medical advice being made universally available.

There is a nightmare scenario. It is that governments license few
people to offer these services, but give them total freedom to use
them as they like. Then, instead of great educational and self-help

institutions – like the public libraries – we are likely to get great influences to exploit and degrade – like video nasties and gutter newspapers.

As people become more aware of the issues involved, it is to be hoped that these issues become more populist and an increasingly important part of every political party's agenda.

So, in the near term, the possibilities on offer in the microelectronics supermarket of the 1990s are just as much at the mercy of the politicians and the regulators as they are of the technologists.

Already, the power of computers to generate images such as video games, and to deliver them to personal computers in the home via phone lines is beginning to worry regulators around the world, who are concerned about the behavioral and addictive effects they may have on children.

Meanwhile, at a rapid rate in the mid-1990s, more and more computer owners log onto the global 'Internet' – a data/telephone network stretching all over the world using telephone lines to which a computer can be attached via a modem and which is a huge source of information and contacts with people.

When the equipment to receive Virtual Reality images in the home becomes affordable to the consumer, so that these images can be sent over services like Internet, the possibilities for expanding the imagination become mind-boggling. And Virtual Reality equipment is already on the chip price learning curve and will be that cheap one day.

So much for the problems that current technology is throwing up for the regulators and politicians of today. Looking further out into the future, technologists are opening up some quite astonishing new capabilities for microelectronics technology.

Perhaps the most extraordinary possibilities lie in the way in which machine-made electrical signals are being connected up to the electrical signals generated by the brain and transmitted by the nervous system.

For instance, by the mid-1990s some 600 people in the UK had undergone operations to implant chips in their heads. All 600 were deaf as a result of damage to the hearing nerves.

The implanted chip contains a radio receiver that can pick up radio signals from an external microphone, a microprocessor that converts the signals into a form understandable by the brain and a transmitter that can send the converted signals along wires to electrodes implanted into whatever remains of the auditory nerves.

The electrical signals from the electrodes then stimulate the nerves into relaying the signals to the brain. The results are said to produce hearing which is 'crude', but better than total deafness.

The imagination can go haywire considering the potential of this kind of capability. For instance, if we can receive a signal sent direct into the brain – which we can do now – then theoretically we should be able to reverse the process and transmit a message directly from the brain to the outside world, without going through the speaking process.

Instead of being spoken, the thought patterns would be fed to a chip, digitized and transmitted to the outside world, where they could be re-transmitted to someone else's brain via a receiver chip implanted in the head. That means telephones become unnecessary!

To be even more fanciful – if we can send and receive speech messages without a telephone then it could be possible to send and receive digitized video signals without a camera or TV. The signals sent from the eye to the brain could be intercepted and transmitted to the outside world.

The signals could then be transmitted to a receiver chip implanted in another person's head from which they are relayed direct into that person's brain. One person's eyes could do the seeing for someone else's brain! This would be pretty bad news for the manufacturers of television sets! We would be able to pick up broadcasts without the need for external equipment. If sight and sound can be transmitted like this then so, probably, will feel, smell, taste and thought.

Microelectronics-based techniques for artificially seeing, smelling, tasting and touching are being developed around the world and are in different stages of sophistication.

For example, in 1994 researchers at the Massachusetts Institute of Technology implanted a chip in a rabbit that can accept an image, process it into electrical signals understandable by the brain and transmit it to the sight cells in the brain. The intention of the MIT researchers is that the technique will one day be sufficiently sensitive to restore sight to the blind.

Another example is a microelectronics-based nose. Two UK companies, Aromascan and Neotronics, have developed artificial noses. They work by using the change in the electrical conductivity of a polymer when the polymer reacts to the airborne molecules of smells. By converting the changes in conductivity into electrical signals, the systems can distinguish between different smells with the same sensitivity as the human nose.

In the Neotronics nose, all the electronics can be incorporated in three pieces of silicon – one chip has the 'sniffing' polymers on one side and the electronics on the other.

On one side, the polymers react to different smells by translating them into tiny voltage changes, and on the other side, the electronics converts the voltage changes – which are analogue signals – into digital signals. The second piece of silicon does the digital processing to distinguish between and identify the signals; the third piece remembers all the smells stored digitally and calls them up when they need to be compared with a smell.

Once these kinds of capabilities are reduced to silicon, they are, of course, on the chip industry learning curve – capable of almost limitless size and cost reduction.

Converting these smells into digital electrical signals means that they can, of course, be stored, processed or transmitted electronically across any distance and reproduced by an equivalent apparatus on the other side of the world. For instance, sitting on a beach in California it would be possible to sniff the smells of Bombay.

As for electronically simulating touch, the sense of how something feels can already be translated into electrical signals, which can then be transmitted to a glove or body suit containing electrical devices to reproduce that feeling in the appropriate area of the body.

Once it is possible to simulate these senses electronically, it becomes possible to transmit the electronic signals across any distance and reproduce them in other people. So the rudimentary technology exists by which you could have the smells, tastes and feelings of one place transmitted to another – in just the same way as we can already transmit the sights and sounds.

At first it will only be possible to receive electronic feelings and Virtual Reality sights using cumbersome equipment with helmets and suits, but the day can be envisaged when the incredible shrinking transistor will make it possible for all those streams of digital data to be delivered straight to an implanted chip in the body, reprocessed into analog form, and fed directly into the brain.

As well as the senses, some researchers think it will be possible to simulate thoughts electronically and transmit them electronically. Thought transference researchers in Japan – at Fujitsu and Hokkaido University – and in America – at the University of

Illinois, at the New York State Health Department and at BioMagnetic Technologies Inc. of San Diego – reckon they can recognize 'brainwaves', i.e. electrical signals generated by the brain, and can distinguish between them.

These researchers have already produced helmets made up of sensors that can 'read' brainwaves and identify the different signals that represent different vowel and word sounds. They are talking about such things as a 'thought-operated' computer, where you transfer your thoughts directly to a computer by radio or wire.

Clearly, once it is possible to recognize brainwaves then it becomes possible to teach machines to recognize the brainwaves and to respond to them. The brain/machine link will become an increasingly important area of development with, literally, mind-boggling implications.

It is one thing to make chips that can replace parts of the body and link them to a human brain, but it is quite another to build an artificial brain into which artificially created sense organs can be linked by an artificial nervous system. However, that is exactly what researchers around the world are working on.

Researchers aim to build silicon brains that can not only 'see' images and pictures but can recognize them for what they are. In a way, it's the 'ultimate' computer that can think, see, hear, smell, taste, see and feel – feel, that is, in the sense of touch; whether or not it will have 'feelings' is anyone's guess!

The building block for making an artificial brain is a silicon chip modelled on the cells, or neurons, of the brain.

Unlike conventional chips, neuron chips have a great many connections to other neurons and, when linked up, resemble a web or a net. Such webs are called 'neural networks', or neural nets.

Neural nets have particular advantages in calculating quickly a vast number of 'If-Then' inferences, i.e 'if the barometer is rising then it will be sunnier' or 'if interest rates are falling then stock prices should rise'.

So computers using neural nets are good at drawing conclusions from very large amounts of changing data inputs, making them useful, for instance, for predicting weather or stock market fluctuations.

Neural nets, being made of silicon chips, are already on the chip industry learning curve and likely to double in power and halve in size roughly every two years. It is almost inevitable that they will reach the same size/power ratio as the human brain – and surpass it.

This, of course, is the route to the ultimate machine – an artificial human being – and the ways to get there are becoming apparent. However, we are all so familiar with androids (through Hollywood) that we will no doubt think it no big trick when we do.

However, microelectronics does appear to be a route to creating an android. If it is ever achieved, the interesting dilemma for all those of us who are carbon-based life-forms is whether we shall opt to become a silicon-based 'life-form' rather than die.

For by the time we can create an android, we will probably have figured out a way to download a human brain into an artificial brain. And so by then the option will exist to transfer from a carbon-based brain to a silicon-based brain. Would we take the option? Would we want immortality?

All of which is interesting fodder for the imagination, but for this century the use of microelectronics to replace human functions inside the body is likely to be confined to relaying externally generated electrical stimuli to damaged nerves, either to get them working again or to partly replace them.

Other possible applications are extremely controversial and derive from implanting chips in animals. For instance, chips implanted in animals' bodies are widely used for 'tagging', i.e. providing a way of finding where they are. They have been implanted in fish and deer to track their whereabouts and are frequently used in cows to provide information on an animal's history.

This has led to criminologists suggesting that chips would be good for implanting in criminals so that they can be tagged – either to check where they are during a term of punishment or, for regular offenders, to provide a record of their movements to check to see if they have been at the scene of a crime.

From there it might be possible to implant chips in the head that can receive an external signal and transmit it directly to the brain in such a way that the brain acts on it. The appalling prospect of radio-controlled people will hopefully never materialize, though it has undoubtedly been a dream of generals and dictators.

If all these ideas on the future directions which microelectronics technology might take seem pretty dreadful to you, remember Xerox. The paperless office that so alarmed the Xerox management at the end of the 1960s simply didn't happen. Today there is ten times more paper in offices than there was in the early 1970s.

Very often the scenarios that are created for future technology development turn out to be wildly inaccurate. For instance, another great myth of the early 1970s was that computer technology linked into telecommunications technology would mean we would all telecommute – work from home. That has, so far at least, turned out to be a pipedream.

The lesson seems to be that if human nature prefers the privacy of a personal piece of paper to read rather than a screen, or prefers working in the company of other people to working in solitude, then the technology will evolve to suit human nature. Let's hope it continues to do so.

We've looked at some of the more fanciful things that advancing microelectronics technology may do in the future. The other side of the same coin is looking at what new things we will be able to do as the cost of microelectronics technology keeps coming down.

One reason why the Xerox management never manufactured the personal computers it developed in the early 1970s was because they would have been too expensive for the ordinary person. It was only when the shrinking transistor brought the cost of the chips down to the level that made a personal computer possible for two or three thousand dollars that the product became a marketable possibility.

So the decline in chip cost is extremely relevant to thinking up new uses for chips. For instance, take a supermarket trolley that you push past the check-out and the check-out instantly prices every item in the trolley and adds it up.

Such a system has been shown to work. It works by having a chip on every product in the supermarket. The chip stores the price. When the trolley passes the check-out a radio beam energizes the chip, which transmits its information to the automatic checkout, which adds up the total value of the goods in the trolley. A credit card 'swipe' then pays the bill.

No more queues at the check-out! The technology to do that has been demonstrated. The only reason it is not up and running in supermarkets is because the chips cost too much. In order to make the system cost-effective, the price of the chips will have to reduce substantially.

That is not too many years away from being a reality, because the chips will need to contain so little information that they will be tiny. And as the silicon wafers on which chips are made get bigger and bigger – eight inches diameter in the mid-1990s – it is possible to envisage getting several tens of thousands of such chips to a wafer.

Since the cost of making a wafer remains stable at a few hundred dollars, it is quite feasible that, in the next few years, the cost per chip will come down to the level where it is cost-effective for it to be included in supermarket packaging – and that's when the checkout person becomes redundant.

So the imagination can always be looking at the future to see when the cost of incorporating a chip in a product is cheap enough to make it practicable to do so. One of the effects of this in the 1990s is probably going to be seen in the increasing use of chips to put speech into products.

It requires about 32 kilobits of digital memory to record a second of speech – which means you need a 1 megabit chip to be able to store a message of half a minute. Since a 1 megabit chip costs just a few dollars in the mid-1990s, and 16 megabit chips offering eight minutes of speech are being made in volume, then the scope for adding speech to many products is growing fast.

We are already seeing greetings cards with a tune or a message preprogrammed onto them and some in which a personal message can be programmed by the sender. What else would be useful? Cooking instructions on food? Planting instructions on seed packets? Spoken advertisements triggered by the turning of a page in a magazine or newspaper? The possibilities are unlimited.

For instance, a good many things that need to be done get overlooked because their need is not visible or obvious. So a chip in a house-plant pot could warn you verbally when its soil got too dry. A chip in a calendar could warn you that it's Auntie Mary's birthday. A chip in the soap powder packet could tell you when it's running low. In such circumstances a verbal warning can be effective when other warnings are overlooked.

The ingenuity of manufacturers will no doubt mean that the chip will become as ubiquitous an adjunct to even the cheapest products as bar-codes became, so that, as well as speech, quite cheap products will be able to have processing capabilities and communications capabilities.

For instance, you could enter your postal code into the chip on a pot-plant and its microprocessor could work out what fertilizers your type of soil needs to make it ideal for the plant.

As the chip gets ever cheaper, it becomes ever more pervasive in finding new uses just as, at the other end of the scale, the increasing power of the latest chips opens up entirely new possibilities. So the limits to the uses for chips are the limits of the imagination.

Which is where you come in. As a citizen, as a buyer or manufacturer of electronics hardware, and as a provider or user of electronics-based services, you hold the ultimate key to the way in which microelectronics is going to be applied – for good or ill.

As a citizen/voter it will be up to you to decide whether you want to see every street corner bugged and on camera; whether you want to see every road electronically 'tolled'; whether you want chips to be implanted in criminals; whether you want governments to have the right to bug all the new communications networks; whether you want governments to delay the new networks by stalling on allocating frequencies and deciding standards; whether you want the supply of pornographic and violent material to be controlled.

As a manufacturer of hardware, it will be your imagination that combines the increasing power and decreasing price of microelectronics to bring out the blockbuster products that will be the basis of new fortunes and new companies.

As a provider of services, it is your imagination that will decide what kind of material the new publishing media – such as video disks and data superhighways – will be delivering.

Creating that material are new businesses in which the core asset is pure imagination. Only imagination can decide how best to mix words, music, film, speech, animation, drawings and photographs to deliver the most effective message.

The products of such businesses will be used to help provide not only entertainment, but every type of service: estate agencies; dating agencies; travel agencies; employment agencies; advice agencies; financial consultants; providers of training and education; and providers of professional services – lawyers, doctors, accountants.

Providing a mixture of words and speech, film and illustrations, and animation and photos should make the messages more accessible and more understandable than ever before. Moreover, the new media types will allow users to dictate their own pace, to ask questions and to receive answers.

However, the most powerful of all our roles in relation to the technology will be as a purchaser of hardware and services. What we pay for will determine what the technology provides. With a technology so limitless, that is power.

Index

Page numbers appearing in **bold** refer to figures.